Springer Series in Molecular Biology

Series Editor: Alexander Rich

Springer Series in Molecular Biology

Series Editor: Alexander Rich

Yeast Genetics
Fundamental and Applied Aspects
J.F.T. Spencer, Dorothy M. Spencer, A.R.W. Smith, eds.

Myxobacteria
Development and Cell Interactions
Eugene Rosenberg, ed.

DNA Methylation
Biochemistry and Biological Significance
Aharon Razin, Howard Cedar, Arthur D. Riggs, eds.

Cooperativity Theory in Biochemistry
Steady-State and Equilibrium Systems
Terrell L. Hill

Molecular Biology of DNA Methylation
Roger L.P. Adams, Roy H. Burdon

Protein Compartmentalization
Arnold W. Strauss, Irving Boime, Günther Kreil, eds.

Structure, Function and Genetics of Ribosomes
Boyd Hardesty and Gisela Kramer

Protein Compartmentalization

Edited by
Arnold W. Strauss   Irving Boime   Günther Kreil

With Contributions by

Barry E. Carlin   Pablo D. Garcia   Glen Hortin
Günther Kreil   Leander Lauffer   John Paul Merlie
Eric N. Olson   Cecil B. Pickett   Vivian Siegel
Claudia A. Telakowski-Hopkins   Peter Walter
Henry C. Wu   Richard Zimmermann

With 26 Figures

Springer-Verlag
New York Berlin Heidelberg Tokyo

Arnold W. Strauss
Department of Biological Chemistry
Washington University School of Medicine
St. Louis, Missouri 63110, U.S.A.

Irving Boime
Department of Pharmacology
Washington University School of Medicine
St. Louis, Missouri 63110, U.S.A.

Günther Kreil
Institute of Molecular Biology
Austrian Academy of Sciences
A-5020 Salzburg
Austria

Series Editor:
Alexander Rich
Department of Biology
Massachusetts Institute of Technology
Cambridge, Massachusetts 02139, U.S.A.

QP
551
.P6958
1986

Library of Congress Cataloging in Publication Data
Protein compartmentalization.
  (Springer series in molecular biology)
  Bibliography: p.
  Includes index.
  1. Proteins—Metabolism. 2. Biological transport.
I. Strauss, Arnold W. II. Boime, Irving. III. Kreil,
Günther. IV.Series.
QP551.P6958 1986   574.87'6042   86-1756

Typeset by Publishers Service, Bozeman, Montana.
Printed and bound by R.R. Donnelley & Sons, Harrisonburg, Virginia.
Printed in the United States of America.

9 8 7 6 5 4 3 2 1

ISBN 0-387-96292-1 Springer-Verlag New York Berlin Heidelberg Tokyo
ISBN 3-540-96292-1 Springer-Verlag Berlin Heidelberg New York Tokyo

Series Preface

During the past few decades we have witnessed an era of remarkable growth in the field of molecular biology. In 1950 very little was known of the chemical constitution of biological systems, the manner in which information was transmitted from one organism to another, or the extent to which the chemical basis of life is unified. The picture today is dramatically different. We have an almost bewildering variety of information detailing many different aspects of life at the molecular level. These great advances have brought with them some breathtaking insights into the molecular mechanisms used by nature for replicating, distributing, and modifying biological information. We have learned a great deal about the chemical and physical nature of the macromolecular nucleic acids and proteins, and the manner in which carbohydrates, lipids, and smaller molecules work together to provide the molecular setting of living systems. It might be said that these few decades have replaced a near vacuum of information with a very large surplus.

It is in the context of this flood of information that this series of monographs on molecular biology has been organized. The idea is to bring together in one place, between the covers of one book, a concise assessment of the state of the subject in a well-defined field. This will enable the reader to get a sense of historical perspective—what is known about the field today—and a description of the frontiers of research where our knowledge is increasing steadily. These monographs are designed to educate, perhaps to entertain, certainly to provide perspective on the growth and development of a field of science which has now come to occupy a central place in all biological phenomena.

The information in this series has value in several perspectives. It provides for a growth in our fundamental understanding of nature and the manner in which living processes utilize chemical materials to carry out a variety of activities. This information is also used in more applied areas. It promises to have significant impact in the biomedical field where an understanding of disease processes at the molecular level may be the capstone which ultimately holds together the arch of clinical research and medical therapy. More recently in the field of biotechnology, there is another type of growth in which this science can be used with immense practical consequences and benefit in a variety of fields ranging from agriculture and chemical manufacture to the production of scarce biological compounds for a variety of application.

This field of science is young in years, but it has already become a mature science. These monographs are meant to clarify segments of this field for the readers.

Cambridge, Massachusetts   Alexander Rich
                                                                    Series Editor

Preface

This monograph addresses a central issue of present day molecular and cellular biology—the mechanisms by which proteins that are synthesized in the cytoplasm ultimately come to reside in other intra- and extracellular compartments or organelles of the cell.

Information regarding the regulation of this "traffic" of newly synthesized proteins as they proceed from their site of synthesis to their ultimate destinations either within or outside the cell has accumulated at a dramatic rate in the last few years. Nearly twenty years ago, several investigators began to elucidate the mechanisms by which secretory proteins were segregated from intracellular proteins through morphological and cell fractionation experiments. The recognition that polysomes engaged in protein synthesis could be separated into two classes, those associated with membranes and those free in the cytoplasm, and that the products of mRNA translation within these two classes of polysomes were secretory and cytoplasmic proteins, respectively, led to the observations of Chefurka and Hayashi and Redman and Sabatini implicating the nascent polypeptide chain as the signal mediating this class separation. Utilizing the experimental evidence of several investigators concerning immunoglobulin and peptide hormone synthesis, a more formal signal hypothesis was presented by Blobel and Dobberstein in 1975 to explain the mechanism by which nascent secretory proteins cross the endoplasmic reticular (ER) membrane during translation of their mRNAs. The concepts of vectorial discharge and cotranslational translocation of nascent precursor polypeptide chains were proposed as general mechanisms for the compartmentalization of proteins destined for many different intracellular organelles (e.g., plasma membrane, Golgi membrane, lysosomes, peroxisomes, mitochondria, and chloroplasts).

This *signal hypothesis* became the impetus for many experiments aimed at testing its validity and at the isolation and characterization of the proposed protein components, including signal peptides, receptors, proteases, and transporters, associated with cotranslational translocation. Similarly, testing of the hypothesis in bacterial and fungal systems to examine protein biosynthesis and compartmentalization became popular. It was soon evident, however, that alternative explanations for some aspects of protein compartmentalization were needed, and other hypotheses were proposed by Wickner and coworkers (the membrane-trigger hypothesis) and Inouye and coworkers (the loop hypothesis). Clear evidence for compartmentalization of mitochondrial, peroxisomal, and chloroplast proteins, and of some bacterial secretory and periplasmic proteins after completion of protein synthesis (posttranslational translocation) emerged. Moreover, specific translocation of both secretory and other organellar proteins without the obligatory, proteolytic removal of a precursor peptide was found. Finally, a role for posttranslational modifications in the compartmentalization of newly synthesized proteins became evident. These alterations included fatty acylation, further proteolytic processing, asparagine-linked glycosylation, processing of oligosaccharide side-chains, acetylation, amidation, disulfide bond formation, and others. Moreover, interactions with other molecules—including receptor proteins, carrier proteins, other protein subunits of multimeric proteins, membrane lipids, and small ribo-nuclear particles—were established as important in protein compartmentalization, particularly during the diversion of lysosomal proteins and integral membrane proteins from the secretory pathway. Nonetheless, many aspects of the signal hypothesis have remained valid and a central role of the ER translocation apparatus in protein compartmentalization has held true.

In the subsequent chapters of this monograph, many aspects of protein compartmentalization will be reviewed and discussed. During the germination of this book, two of the editors (A. W. Strauss and I. Boime), discussed how to best present the information to the reader. Due to the large amount of evidence reviewed, the varying viewpoints concerning mechanism of compartmentalization, and the rapid progress in the field, we asked several authors to prepare chapters on specialized topics. We believed that this approach would allow the expression of a broad range of opinions concerning the interpretation of the experimental data and also provide the best method of expert and detailed presentation of many aspects of the problem. In view of the major contributions of European investigators to the field, a European editor (G. Kreil) was recruited. We hope that the strengths of the multiauthor approach are evident during reading of the subsequent chapters. Unfortunately, this approach, of necessity, allowed for some repetition of material and created some difficulties in generating simultaneous submission of manuscripts and delay in publication. Consequently, the text reflects the thoughts of the authors and the status of the literature as current as early 1985. Moreover, some aspects of protein compartmentalization are not discussed. These include oligosaccharide addition and processing and lysosomal and chloroplast protein biosynthesis. Such limitations are the faults of the editors. However, we hope that the material presented will provide the reader with a strong background in the area of protein compartmentalization and an appreciation for the many unresolved problems in the field.

Chapters 1–3 deal with the signal hypothesis and cotranslational translocation of secretory proteins. In the initial chapter, Peter Walter and his colleagues review the status of the signal hypothesis and the role of SRP in the translocation of secretory and lysosomal proteins across the ER membrane and the integration of membrane proteins into the ER membrane. In the second chapter, Glen Hortin discusses some potential limitations of the signal hypothesis as currently understood and proposes a modification of it entitled the "Allosteric Model" for recognition of signal peptides. In the third chapter, Henry Wu reviews the structure of signal peptides and the structural requirements for their binding to SRP and for their proteolytic removal by signal peptidases.

Chapters 4–6 concern three posttranslational modifications that proteins undergo within the compartments of the secretory pathway which are critical in the normal biosynthesis and activation of both secretory and integral membrane proteins. In the fourth chapter, Günther Kreil reviews the posttranslational processing of pro-protein precursors. Dr. Kreil discusses the generation from a single large precursor of multiple, smaller active peptides and the unusual proteolytic processing through sequential removal of dipeptides which occurs posttranslationally. In Chapter 5, John Merlie and Barry Carlin present evidence for the role of subunit interactions in the regulation of biosynthesis and assembly of multimeric proteins. The importance of these interactions, as they occur within the various compartments of the secretory pathway, in determining subsequent modifications of the peptide backbone, and in prevention of proteolytic degradation of constitutive subunits has only recently been appreciated. Many aspects of the biosynthesis of plasma membrane proteins are discussed in this chapter. In the sixth chapter, Eric Olson reviews the covalent addition of fatty acids to the peptide backbone and the role that this modification has in the anchoring or segregation of intracellular membrane proteins. While the occurrence of this modification in bacteria and brain proteolipoprotein was described many years ago, the potential importance of fatty acid acylation in protein compartmentalization within other mammalian cells has been only recently suggested.

The final chapters describe compartmentalization of microsomal and mitochondrial proteins. Study of the biosynthesis and processing of these proteins has, until the last five years, been hampered by difficulties in protein purification and in determination of primary structure. Cecil Pickett and Claudia Telakowski-Hopkins review the biosynthesis of some integral membrane proteins of the ER and Golgi apparatus. While most of these proteins utilize the ER translocation apparatus, the authors describe significant differences. The biosynthesis of mitochondrial proteins is reviewed by Richard Zimmermann, who with Walter Neupert and Schatz and coworkers exemplifies the major contributions of European workers in this area. The compartmentalization of mitochondrial proteins synthesized from mRNA derived from the nuclear genome in the cytoplasm is both the most thoroughly studied example of posttranslational uptake of proteins by organelles and one of the most rapidly expanding current fields of investigation.

All of the authors have emphasized the many questions and areas for investigation that remain concerning protein compartmentalization. These include, for example, characterization and purification of a eukaryotic signal peptidase; the role of ribosomal proteins and RNA in SRP interactions with signal peptides; characterization of the SRP receptor protein ("docking protein"); the mechanism that allows translocation of the peptide backbone through the ER membrane; the substrate within the peptide backbone that determines differential processing of oligosaccharides and fatty acid acylation; the role of receptors in the ER in diversion of proteins to various pathways; the role of pH of intracellular compartments in secretory pathway traffic; and characterization of pro-protein peptidases. Further, the mechanisms by which translocation across the ER membrane of integral membrane proteins is arrested during their synthesis and how these proteins associate with other proteins required for their activity and are correctly transported within membranes to their ultimate destinations are only beginning to be investigated. Virtually nothing is known, as yet, concerning the proteins of the mitochondrial translocation apparatus, nor even how many different mitochondrial protein uptake pathways exist. Currently, application of the techniques of molecular biology and cell biology, particularly through the creation of site-specific and deletion mutations, is beginning to generate answers to many of these questions.

Arnold W. Strauss
Irving Boime
Günther Kreil

Contents

Contributors

BARRY E. CARLIN Department of Pharmacology, Washington University, St. Louis, Missouri 63110, U.S.A.

PABLO D. GARCIA Department of Biochemistry and Biophysics, University of California, San Francisco, California 94143, U.S.A.

GLEN HORTIN Division of Laboratory Medicine, Washington University School of Medicine, St. Louis, Missouri 63110, U.S.A.

GÜNTHER KREIL Institute of Molecular Biology, Austrian Academy of Sciences, A-5020 Salzburg, Austria

LEANDER LAUFFER Department of Biochemistry and Biophysics, University of California, San Francisco, California 94143, U.S.A.

JOHN PAUL MERLIE Department of Pharmacology, Washington University, St. Louis, Missouri 63110, U.S.A.

ERIC N. OLSON Anderson Hospital and Tumor Institute at Houston, Houston, Texas 77030, U.S.A.

CECIL B. PICKETT Department of Molecular Pharmacology and Biochemistry, Merck Sharp & Dohme Research Laboratories, Rahway, New Jersey 07065, U.S.A.

VIVIAN SIEGEL Department of Biochemistry and Biophysics, University of California, San Francisco, California 94143, U.S.A.

CLAUDIA A. TELAKOWSKI-HOPKINS Department of Molecular Pharmacology and Biochemistry, Merck Sharp & Dohme Research Laboratories, Rahway, New Jersey 07065, U.S.A.

PETER WALTER Department of Biochemistry and Biophysics, University of California, San Francisco, California 94143, U.S.A.

HENRY C. WU Department of Microbiology, Uniformed Services University of the Health Sciences, Bethesda, Maryland 20814, U.S.A.

RICHARD ZIMMERMANN Institut für Physiologische Chemie, Physikalische Biochemie und Zellbiologie, Ludwig-Maximilians-Universität München, Federal Republic of Germany

1

The Protein Translocation Machinery of the Endoplasmic Reticulum: The Signal Hypothesis Ten Years Later

PETER WALTER, VIVIAN SIEGEL, LEANDER LAUFFER, and PABLO D. GARCIA

Abstract

In 1975, the signal hypothesis was proposed by Blobel and Dobberstein to explain the targeting and translocation of secretory proteins across the endoplasmic reticulum (ER) membrane. Ten years later, molecular components of the cellular machinery catalyzing these events have been isolated and characterized. Our current view of the function of the signal recognition particle (SRP) and the SRP receptor (or docking protein) in the translocation of secretory and lysosomal proteins across—as well as integration of integral membrane proteins into—the ER membrane is described. In brief, SRP is thought to recognize the signal peptide on the nascent proteins and to arrest their translation in the cytoplasmic space. Upon interaction of the elongation-arrested ribosome with the correct target membrane (the ER), and particularly after a direct interaction of the ribosome-bound SRP with the SRP receptor in the ER membrane, the elongation arrest is released. A functional ribosome-membrane junction is established, allowing the translocation of the nascent polypeptide across the membrane by an as yet poorly understood mechanism.

In the last few years, major advances have been made in our understanding of the targeting and translocation machinery of the ER. Three distinct classes of proteins are known to utilize this translocation system: secretory (1) and lysosomal (2) proteins are translocated across the ER membrane, while certain classes of integral membrane proteins (3) are integrated into it. A detailed biochemical analysis of this process became feasible through the development of *in vitro* systems that were able to reproduce the translocation of nascent secretory proteins across the ER membrane (isolated in the form of closed microsomal vesicles) with apparent fidelity (4, 5). So far, two components have been purified from canine pancreas and shown to be required for the translocation event. We will introduce the reader to the biochemistry of these components and discuss their role in protein translocation.

Structure of SRP

One of these components is the signal recognition particle (SRP), which is an 11S cytoplasmic ribonucleoprotein. SRP restores the translocation activity of salt-extracted microsomes *in vitro* (54), and it was purified from a salt extract of pancreatic microsomal vesicles using this activity assay (6). SRP consists of six nonidentical polypeptide chains that are organized in four SRP proteins (two monomers composed of a 19-kd polypeptide and a 54-kd polypeptide, and two heterodimers: one composed of 9-kd and 14-kd polypeptides, and the other composed of 68-kd and 72-kd polypeptides, respectively) (6, 7) and one molecule of RNA of about 300 nucleotides (8). When separated under native conditions, both the RNA and protein fractions are inactive by themselves, but together, they can be readily reconstituted into an active particle (9).

The SRP-RNA was first discovered as a component of oncornaviruses (10), but it was later shown to be a constitutive component of all noninfected cells (11). It is not known whether the RNA plays any functional role in the virus particles. It was variously localized to cytoplasmic, ribosomal, microsomal, and nuclear cell fractions. In retrospect, many of these discrepancies can be reconciled by the fact that SRP partitions into various cell fractions, depending upon the precise ionic conditions of the homogenate (12). The different names given to the RNA (small cytoplasmic RNA "L," 7S RNA) were combined by Ullu and coworkers to "7SL RNA" (13). Sequence analysis (13, 14) confirmed the previous observation (15) that long stretches of 7SL RNA are homologous to the Alu family of highly repetitive DNA sequences in the higher mammalian genome. 7SL RNA was shown to be about 80% homologous to an Alu consensus sequence for about 100 nucleotides at its 5' end and 50 nucleotides at its 3' end. The core portion (termed the S sequence, about 150 nucleotides long) showed no homology to Alu DNA, but it is present in the genome at a middle repetitive frequency. Most of these copies, however, are likely to represent transcriptionally silent pseudogenes, leaving only a few active 7SL RNA genes per mammalian genome (E. Ullu, unpublished).

Little is known about the function of 7SL RNA in SRP, besides the fact that it acts as a structural backbone in the particle. In the absence of 7SL RNA, SRP proteins behave as monomers or heterodimers with little or no affinity for each other (7, 9). Nucleolytic digestion of SRP causes its inactivation and disintegration (8). A limited nucleolytic digestion appears to sever SRP into two "domains"; one containing mainly S sequences with the 72-, 68-, 54-, and 19-kd polypeptides attached, and the other one containing both 3' and 5' Alu sequences with the 14- and 9-kd polypeptides bound (16). Thus, the dipartite structure of 7SL RNA (Alu-S-Alu) may also be reflected in intact SRP, since 7SL RNA

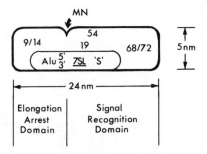

Fig. 1.1. Schematic presentation of SRP structure. The arrow indicates where micrococcal nuclease (MN) severs SRP into two subparticles. The positions of the indicated polypeptides within a subparticle are arbitrary.

appears to fold back onto itself, thereby allowing the 5′ and 3′ Alu segments to base pair with each other (17). It is likely that these structural domains also reflect functional domains in SRP (see below). In Figure 1.1, we show a schematic representation of SRP reflecting our current working model.

A variety of electron microscopic (EM) techniques have been applied to directly visualize SRP (18). Images were obtained either after negative staining with or without fixation or after platinum shadowing. Also, unstained specimens were examined by darkfield electron microscopy. All images are internally consistent and show SRP as a rod-like structure of about 24-nm length and 5-nm width (see also Figure 1.1). From these dimensions, we calculate a volume of 525 nm³, assuming a perfect cylindrical shape, which is in good agreement with the calculated volume based on SRP's molecular composition (the combined molecular weight of RNA and protein components is 325 kd, resulting in an expected volume of 380 nm³) if one takes into account that various crevices and indentations are reproducibly seen in the images with SRP therefore deviating considerably from perfect rod shape (18).

Signal Recognition and Elongation Arrest

The function of SRP (and SRP receptor) in the protein translocation process was deduced using *in vitro*-assembled polysomes that were programmed with mRNAs coding for specific secretory proteins. Thus, it became possible to directly follow the fate of a nascent secretory protein, and thereby to analyze the discrete steps of its recognition and subsequent translocation across the membrane of microsomal vesicles. The data accumulated thus far are discussed in the model depicted in Figure 1.2 (29), which represents an "updated" version of the signal hypothesis as originally formulated by Blobel and Dobberstein (4).

It was shown that free (soluble) SRP (a in Figure 1.2) can exist in equilibrium with a membrane-bound form (bound presumably to SRP receptor (e in Figure

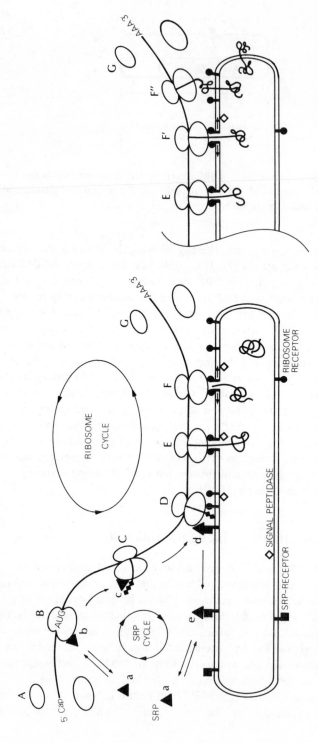

Fig. 1.2. Model for protein translocation across the ER membrane. (Reproduced with permission from [29]. Copyright M.I.T.)

1.2), as well as with a ribosome-bound form (b in Figure 1.2) (12). Upon translation of an mRNA coding for a signal sequence that targets it to the ER translocation system (A-C in Figure 1.2), the apparent affinity of SRP for the translating ribosome was enhanced by three to four orders of magnitude (C in Figure 1.2) (23). Thus, SRP could be shown to *recognize* the signal sequences of nascent secretory (23, 30, 31) and lysosomal proteins (32), as well as membrane (33) proteins.

Interestingly, concomitant with this increase in binding affinity, SRP specifically *arrested* the elongation of the initiated polypeptide chain at a discrete site (just after the signal peptide had emerged from the ribosome), thereby preventing completion of the presecretory protein (C in Figure 1.2) (24, 27). The mechanism by which SRP interacts with the ribosome and signal sequences to cause recognition and elongation arrest is unknown. However, the detection of signal sequences by SRP clearly involves the recognition of the nascent polypeptide chain *per se*, as distinguished from recognition of the mRNA template, because a perturbation of the nascent polypeptide through the incorporation of certain amino acid analogs renders signal sequences nonrecognizable (23, 34). It still remains a puzzle as to what the recognized consensus feature of signal sequences is. Since signal sequences lack primary sequence homology (35), SRP must recognize certain features contained in the secondary or tertiary structure of these peptides. It is unknown whether the increase in affinity of SRP to the ribosome results from a conformational change in SRP (or the ribosome) induced by the emergence of the nascent signal peptide and/or from a direct interaction of SRP with the signal sequence itself. Also, while it is an intriguing possibility that 7SL RNA, as part of SRP, is somehow involved in a direct interaction with other RNA species of the translation apparatus (rRNA, mRNA, possibly even tRNA), there is currently no experimental evidence that such a mechanism plays a role in SRP function.

The knowledge about the molecular dimensions of SRP (18) has considerably influenced our view of how SRP could functionally interact with the protein translation machinery. While the molecular details of the elongation arrest reaction still remain nebulous, they raise the question as to how SRP could recognize information in the nascent chain and simultaneously affect reactions involved in elongation, since the nascent chain exit site (36) and the peptidyl transferase center (37) in the ribosome are believed to be physically separated by about 16 nm (Figure 1.3) (36). While it is, of course, possible that such a mechanism involves allosteric changes across the ribosome itself, it is also conceivable that SRP physically bridges the distance between the two sites (Figure 1.3), thus recognizing (binding to) signal sequences with one end and modulating the elongation reaction with the other (possibly by blocking the aminoacyl-tRNA binding site or by directly interfering with the peptidyl transferase).

Recently, we were able to demonstrate that the elongation arrest reaction is not a prerequisite for translocation (7). This was demonstrated by using an SRP that was reconstituted from separate protein and RNA components and from which the 9/14-kd SRP protein was omitted [SRP(−9/14)]. We discovered that

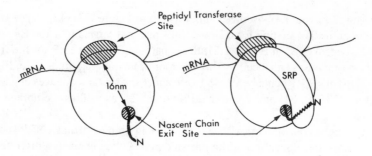

Fig. 1.3. Current working model of the functional binding of SRP to the ribosome.

SRP($-$9/14) was active in the translocation assay, although at a reduced level relative to that of the complete particle. Surprisingly, this particle inhibited neither the rate nor the degree of secretory protein synthesis; the decreased translocation activity of SRP($-$9/14) could be fully accounted for by the observation that without elongation arrest, the functional interaction of the polysome with the microsomal membrane had acquired a strict time dependence. If elongation were allowed to proceed too far in the presence of SRP($-$9/14) before the addition of microsomal membranes, translocation was no longer observed. In addition, one could more closely approach the translocation activity of complete SRP simply by increasing the membrane concentration. We concluded from this result that we had not, in fact, affected the translocation activity of the particle by removing these two polypeptides. In other words, we have completely abolished one of the assayable activities of SRP (the elongation arrest) without noticeably altering the other.

It seems reasonable that the portion of SRP comprised of the 9/14-kd protein, and the RNA it binds to, is contained in a structural domain separate from the rest of the particle based on the dissection of SRP by limited nucleolytic digestion (16) described above. Since, according to these experiments, it is the Alu-like portion of 7SL RNA that interacts with the 9/14-kd SRP protein, which in turn is shown to be responsible for the elongation arrest activity of SRP, the Alu sequences in SRP are put in direct context with its elongation arrest activity (see Figure 1.1). The provocative possibility therefore arises that Alu transcripts, in general, may function in some aspect of translational control.

Given that the elongation arrest is not absolutely required for protein translocation *in vitro*, the question about its function and importance *in vivo* becomes an even more interesting one. Elongation arrest clearly could serve a fidelity function by 1) preventing the synthesis of presecretory and prelysosomal proteins in the cytoplasmic compartment and 2) retaining the nascent chain in a translocation-competent state. However, elongation arrest could also be exploited as a regulatory step, providing the cell with a fast, possibly selective, on-off switch modulating specific secretory or membrane protein synthesis at the level of elongation. This could be achieved either by other, as yet undefined, components or by direct modification of SRP and/or SRP receptor.

Structure of SRP Receptor

Using the same *in vitro* protein translocation assays as the principle tool that led to the purification of SRP, two distinct approaches were taken to identify membrane components involved in this process. Both approaches eventually converged in the discovery and purification of the SRP receptor as the first—and so far only—integral membrane protein that has been proven to play an essential role in this process.

One of these approaches was based on the early observation (19) that proteolysis of microsomal membranes completely abolished their protein translocation activity, but that—most importantly—it could be restored upon readdition of an extract prepared by limited proteolysis of the starting microsomal membrane fraction. This proteolytic dissection and functional reconstitution provided the assay for the purification of the protease-solubilized component. Meyer and coworkers purified this activity as a basic 60-kd protein (20, 21), and they demonstrated through the use of immunologic techniques that it was indeed a proteolytic fragment and that it was derived from a 72-kd integral membrane protein of the ER (22).

The second approach took advantage of the observation that when assayed in the absence of microsomal membranes, SRP caused a site-specific elongation arrest in the synthesis of presecretory proteins (23), and that microsomal membranes contained an activity capable of releasing the elongation arrest (24). Most important, this arrest-releasing activity was still active in a detergent extract of the microsomal membrane fraction (25). Fractionation of the detergent extract—employing affinity chromatography on SRP-Sepharose as a key step—allowed the purification of a 72-kd integral membrane protein that contained the arrest-releasing activity (26). Using both immunologic and peptide-mapping techniques, this protein was shown to be identical to the 72-kd protein identified via the proteolytic dissection route described above, and was termed "SRP receptor" (26) (or "docking protein" by the Heidelberg group [27]).

Recently, the primary structure of the SRP receptor was determined (28). This was accomplished by first determining the first 27 amino acids of purified SRP receptor by using sequential Edman degradations. Using this sequence information, an oligonucleotide was designed and used to probe several cDNA libraries. Five overlapping cDNA clones were isolated which cover the complete coding and 3′-untranslated sequence of the receptor. The amino terminus of the cytoplasmic SRP receptor fragment was shown to begin with residue 152 of the intact SRP receptor. Thus, it is sequences contained within the amino terminal 151 amino acids that anchor the SRP receptor in the lipid bilayer. Indeed, two distinctly hydrophobic regions have been identified that constitute putative transmembrane segments (28). The remainder of the sequence is surprisingly hydrophilic. In particular, clusters of charged residues are found surrounding the proteolytic cleavage site that may be involved in nucleic acid binding. The possible functional significance of these regions is discussed below.

Targeting and Release of Elongation Arrest

Upon interaction of the SRP-arrested ribosomes with microsomal membrane (D in Figure 1.2), the elongation arrest was *released* (24). This arrest-releasing activity of the microsomal membrane fraction was localized to the SRP receptor (26). Since SRP receptor is purified by virtue of its affinity to SRP itself, it is likely that it is also the direct interaction between SRP on the arrested ribosome and SRP receptor that is responsible for the release of the arrest (D in Figure 1.2). Interestingly, while purified SRP receptor in detergent solution possessed arrest-releasing activity (26), the cytoplasmic proteolytic fragment of the receptor did not (7, 25). A conflicting report (27) to this last point can be reconciled, because the preparations of the proteolytic receptor fragment used in these studies were shown to also contain some intact receptor (22, 26). In addition to the absence of arrest-releasing activity, the 60-kd cytoplasmic SRP receptor fragment also did not show any affinity for SRP (28).

These observations can be interpreted in light of the structural features of the SRP receptor. In particular, it is striking that the regions flanking these protease cleavage sites are dominated by a remarkable abundance of charged amino acids that mostly appear in clusters (28). These clusters resemble sequences in ribosomal proteins and aminoacyl-tRNA synthetases that were implied to be involved in binding the proteins to RNA molecules. This finding raised the interesting possibility that the SRP receptor could interact directly through this charged domain with the 7SL RNA in SRP, thus becoming transiently an integral part of the particle. Since both the membrane anchor fragment and the cytoplasmic fragment contributed jointly to this domain, its separation by proteolytic cleavage could account for the apparent lack of activity of both SRP receptor fragments.

Preliminary quantitations indicate that in pancreatic cells, both SRP and SRP receptor are present only in substoichiometric amounts with respect to membrane-bound ribosomes (26). This suggests that the ribosome-SRP-SRP receptor interaction might be a transient one, merely targeting the SRP-arrested ribosome to a specific translocation-competent site on the ER membrane. This conjecture is supported by recent data demonstrating that purified SRP receptor in detergent solution—concomitant with releasing the elongation arrest—causes SRP to lose its high affinity to the signal-bearing ribosome (38). Since SRP receptor itself showed no measurable affinity to ribosomes, it appears that both SRP and SRP receptor are free to be recycled once the correct targeting of the ribosome to the ER membrane has been accomplished ("SRP-cycle," a-e in Figure 1.2). Furthermore, the SRP receptor is required for protein translocation, even in the absence of elongation arrest (catalyzed by SRP[−9/14] as described above) (7). Such an absolute requirement for SRP receptor might be reflective merely of the affinity between SRP and SRP receptor, *i.e.*, SRP receptor is solely required to correctly target the ribosome to the ER membrane. Alternatively, SRP receptor might be involved in the initiation of the translocation process itself, either directly or by organizing in its proximity whatever components are required for protein translocation to occur.

Protein Translocation

The steps following this initial targeting event are presently only poorly understood. Once targeting has occurred, the ribosome-SRP-SRP receptor interaction is most likely replaced by a direct interaction of the ribosome with other integral membrane proteins (here indicated as putative ribosome receptors), leading to the formation of a functional ribosome membrane junction (D-F in Figure 1.2). Two integral membrane proteins (ribophorin I and II) that might function as ribosome receptors have been described by Kreibich and coworkers (39, 40), and they appear to be involved in ribosome binding to the ER membrane and are present in the membrane in stoichiometric proportions to membrane-bound ribosomes (54). However, their direct involvement in protein translocation has not yet been demonstrated and has been questioned by other investigators (41).

The actual *protein translocation* has been depicted to occur via the formation of a "pore-like" structure (E in Figure 1.2), solely to indicate that the permeability barrier of the membrane is somehow reversibly disrupted to allow the passage of the nascent polypeptide. While it was argued on thermodynamic grounds (42) that protein translocation could occur without the involvement of membrane proteins, there is evidence to suggest that additional proteins may, in fact, be required (25). However, whether translocation actually occurs through a proteinaceous, water-filled channel or through some "changed environment" that causes the interior of the bilayer to be more amphiphilic, or even directly through the lipid bilayer, remains to be experimentally addressed.

During translation/translocation, a variety of enzymatic activities localized on the luminal side of the ER act on the nascent polypeptide. The signal peptides are removed (in most cases) by signal peptidase, high mannose core oligosaccharides are transferred to selected asparagine residues, and cysteine residues are oxidized to form disulfide bonds. Neither cleavage by signal peptidase (43) nor core glycosylation (44) are required for productive translocation to occur. The enzymes catalyzing these reactions, therefore, may not be essential components of the translocation machinery. However, the knowledge of the temporal occurrence of these modifications, with respect to chain elongation, provides a convenient measure of the membrane-spanning portion of the nascent chain. Thus, it was possible to measure the length of the nascent chain between the peptidyl transferase site in the ribosome and the earliest occurrence either of glycosylation (on ovalbumin) (45) or of disulfide bond formation (on Ig light chain) (46).

These data indicate that the nascent polypeptide must cross the membrane in a rather extended fashion (E in Figure 1.2), and are inconsistent with the notion that any protein folding is required on the cytoplasmic side prior to translocation. Thus, in principle, secretory proteins need not be intrinsically different from cytoplasmic proteins to be compatible with the translocation apparatus. This conjecture is supported by recent data by Lingappa and coworkers (47). These authors created a fusion protein of globin and the amino terminal signal sequence of β-lactamase, thereby demonstrating that, in principle, it is possible to translocate a cytoplasmic protein (globin) across the ER membrane.

In contrast to protein import into mitochondria (recently reviewed in [48,49]), translocation across the ER membrane does not appear to require an electrical potential or an ion gradient, since a variety of uncouplers and ionophores have no effect on the event *in vitro*. Because the in vitro translocation assays are dependent on ongoing protein synthesis, nucleotide triphosphates cannot be omitted; therefore, one cannot rule out their direct involvement in the translocation process at present.

Termination of Translocation and Stop-Transfer

Upon termination of protein synthesis, the completed polypeptide is *released* into the lumen of the ER (F in Figure 1.2). The permeability barrier of the membrane is restored (indicated here by the disassembly of the pore-like structure) and the ribosome is released from the membrane (G in Figure 1.2). It is then free to enter the soluble pool and complete the "ribosome cycle" (A-G in Figure 1.2). In the case of integral membrane proteins, the translocation process is aborted prior to completion of the protein (F' in Figure 1.2), resulting in its asymmetric integration into the membrane (Figure F'' in Figure 1.2). The information causing this arrest of translocation (a putative "stop-transfer sequence") is most likely contained in the nascent chain and interpreted by the translocation machinery. Some translocated proteins (such as the extracellular domain of sendai virus F protein [50]) contain long stretches of hydrophobic amino acids; thus it is not merely hydrophobicity that causes integral membrane proteins to become "stuck" in the lipid biplayer. It is not known whether a ribosome remains membrane-bound or (as indicated here in F'' in Figure 1.2) detaches from the membrane while synthesizing the cytoplasmic domain of a membrane protein.

Most current models imply that the nascent chain actually crosses the membrane in a loop configuration as shown in Figure 1.4. Depending on whether the amino terminus is retained on the cytoplasmic side of the ER (which can be variable, depending on the localization of stop-transfer sequences and the presence or absence of a signal peptidase cleavage site), the amino terminus of the resulting membrane protein will face either the cytoplasm (Figure 1.4) or the lumen of the ER (Figure 1.4). Once this initial asymmetric orientation has been achieved, the insertion of further membrane-spanning segments either by simple partitioning of hydrophobic stretches into the membrane or by a mechanism involving the

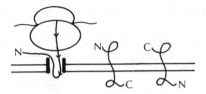

Fig. 1.4. Model for translocation accounting for the asymmetric integration of membrane proteins in both orientations.

reengagement of the translocation machinery (via multiple internal signal sequences [51]) can be envisioned.

Evolutionary Considerations

In closing, it should be noted that both SRP (and its 7SL RNA [52,53]) and the mode of cotranslational protein translocation appear to be highly conserved through evolution. It is possible to construct active chimeric SRPs consisting of mammalian polypeptides and either amphibian or insect 7SL RNA (9). Furthermore, it was demonstrated (31) (P. G. Garcia *et al.*, in preparation) that the signal sequence of nascent *prokaryotic* secretory proteins (β-lactamase or lipoprotein) that were synthesized on *plant* ribosomes (wheat germ) were correctly recognized by *mammalian* SRP (canine); *i.e.*, elongation was arrested, translocation across mammalian membranes was SRP-dependent, and mammalian signal peptidase cleaved the signal peptide of β-lactamase in the correct position. SRP and SRP receptor, therefore, are integral and indispensable constituents of the protein synthesis machinery of living cells. They assure the correct transport of a specific subset of newly synthesized proteins, designated by ER-directed signal sequences, to the correct subcellular location. Considering SRP's structural features and its intimate (albeit transient) functional association with the ribosome, it could almost be regarded as a "third ribosomal subunit" that functions as the adapter between the cytoplasmic translation and the membrane-bound translocation machinery.

Acknowledgments. This work was supported by NIH grant GM 32384. Peter Walter is a recipient of support from The Chicago Community Trust/Searle Scholars Program.

References

1. Palade, G. (1975) Science *189*, 347–358.
2. Erickson, A. H., Conner, G. & Blobel, G. J. (1981) J. Biol. Chem. *256*, 11224–11231.

3. Lingappa, V. R., Katz, F. N., Lodish, H. F. & Blobel, G. (1978) J. Biol. Chem. *253*, 8667–8670.
4. Blobel, G. & Dobberstein, B. (1975) J. Cell Biol. *67*, 852–862.
5. Szczesna, E. & Boime, I. (1976) Proc. Natl. Acad. Sci. USA *73*, 1179–1183.
6. Walter, P. & Blobel, G. (1980) Proc. Natl. Acad. Sci. USA *77*, 7112–7116.
7. Siegel, V. & Walter, P. (1985) J. Cell Biol. *100*, 1913–1921.
8. Walter, P. & Blobel, G. (1982) Nature *299*, 691–698.
9. Walter, P. & Blobel, G. (1983) Cell *34*, 525–533.
10. Bishop, J. M., Levinson, W., Sullivan, D., Farrshier, L., Quintrell, N. & Jackson, J. (1970) Virology *42*, 927–937.
11. Erikson, E., Erikson, R. L., Henry, B. & Pace, N. R. (1973) Virology *53*, 40–46.
12. Walter, P. & Blobel, G. (1983) J. Cell Biol. *97*, 1693–1699.
13. Ullu, E., Murphy, S. & Melli, M. (1982) Cell *29*, 195–202.
14. Li, W. Y., Reddy, R., Henning, D., Epstein, P. & Bush, H. (1982) J. Biol. Chem. *257*, 5136–5142.
15. Weiner, A. (1980) Cell *22*, 209–218.
16. Gundelfinger, E. D., Krause, E., Melli, M. & Dobberstein, B. (1983) Nucl. Acids Res. *11*, 7363–7374.
17. Gundelfinger, E. D., Carlo, M. D., Zopf, D. & Melli, M. (1984) EMBO J. *3*, 2325–2332.
18. Andrews, D. W., Walter, P. & Ottensmeyer, P. (1985) Proc. Natl. Acad. Sci. USA *82*, 785–789.
19. Walter, P., Jackson, R. C., Marcus, M. M., Lingappa, V. R. & Blobel, G. (1979) Proc. Natl. Acad. Sci. USA *76*, 1795–1799.
20. Meyer, D. I. & Dobberstein, B. (1980) J. Cell Biol. *87*, 498–502.
21. Meyer, D. I. & Dobberstein, B. (1980) J. Cell Biol. *87*, 503–508.
22. Meyer, D. I., Louvard, D. & Dobberstein, B. (1982) J. Cell Biol. *92*, 579–583.
23. Walter, P., Ibrahimi, I. & Blobel, G. (1981) J. Cell Biol. *91*, 545–550.
24. Walter, P. & Blobel, G. (1981) J. Cell Biol. *95*, 557–561.
25. Gilmore, R., Blobel, G. & Walter, P. (1982) J. Cell Biol. *95*, 463–469.
26. Gilmore, R., Walter, P. & Blobel, G. (1982) J. Cell Biol. *95*, 470–477.
27. Meyer, D. I., Krause, E. & Dobberstein, B. (1982) Nature *297*, 647–650.
28. Lauffer, L., Garcia, P. D., Harkins, R., Coussins, L., Ullrich, A. & Walter, P. (1985) Nature *318*, 334–338.
29. Walter, P., Gilmore, R. & Blobel, G. (1984) Cell *38*, 5–8.
30. Stoffel, W., Blobel, G. & Walter, P. (1981) Eur. J. Biochem. *120*, 519–522.
31. Muller, M., Ibrahimi, I., Chang, C. N., Walter, P. & Blobel, G. (1982) J. Biol. Chem. *257*, 11860–11863.
32. Erickson, A. H., Walter, P. & Blobel, G. (1983) Biochem. Biophys. Res. Com. *115*, 275–280.
33. Anderson, D. J., Walter, P. & Blobel, G. (1982) J. Cell Biol. *93*, 501–506.
34. Hortin, G. & Boime, I. (1980) Proc. Natl. Acad. Sci. USA *77*, 1356–1360.
35. VonHeijne, G. (1983) Eur. J. Biochem. *133*, 17–21.
36. Bernabeau, C., Tobin, E. M., Fowler, A., Zabin, I. & Lake, J. A. (1983) J. Cell Biol. *96*, 1471–1474.
37. Lake, J. A. & Strycharz, W. A. (1981) J. Mol. Biol. *153*, 979–992.
38. Gilmore, R. & Blobel, G. (1983)Cell *35*, 677–685.
39. Kreibich, G., Freienstein, C. M., Pereyra, B. N., Ulrich, B. L. & Sabatini, D. D. (1978) J. Cell Biol. *77*, 488–506.

40. Marcantonio, E. E., Amar-Costesec, A. & Kreibich, G. (1984) J. Cell Biol. *99*, 2254–2259.
41. Bielinska, M., Rogers, G., Rucinsky, T. & Boime, I. (1979) Proc. Natl. Acad. Sci. USA *76*, 6152–6156.
42. Engelman, D. M. & Steitz, T. A. (1981) Cell *23*, 411–422.
43. Hortin, G. & Boime, I. (1981) Cell *24*, 453–461.
44. Rothman, J. E., Katz, F. N. & Lodish, H. F. (1978) Cell *15*, 1447–1454.
45. Glabe, C. G., Hanover, J. A. & Lennarz, W. J. (1980) J. Biol. Chem. *255*, 9236–9242.
46. Bergman, L. W. & Kuehl, W. M. (1979) J. Biol. Chem. *254*, 8869–8876.
47. Lingappa, V. R., Shaidez, J., Yost, C. S. & Hedgpeth, J. (1984) Proc. Natl. Acad. Sci. USA *81*, 456–460.
48. Hay, R., Bohni, P. & Gasser, S. (1984) Biochim. Biophys. Acta *779*, 65–87.
49. Schatz, G. & Butow, R. A. (1983) Cell *32*, 316–318.
50. Gething, M. J., White, J. M. & Waterfield, M. D. (1978) Proc. Natl. Acad. Sci. USA *75*, 2737–2740.
51. Blobel, G. (1980) Proc. Natl. Acad. Sci. USA *77*, 1496–1500.
52. Ullu, E. & Melli, M. (1982) Nucl. Acids Res. *10*, 2209–2223.
53. Ullu, E. & Tschudi, C. (1984) Nature *312*, 171–172.
54. Warren, G. & Dobberstein, B. (1978) Nature *273*, 569–571.

2

Mechanism and Structural Basis for Recognition of Signal Peptides

GLEN HORTIN

Abstract

The Signal Hypothesis proposes that information for segregation of secretory proteins from cytoplasmic proteins is encoded within signal peptides by a highly specific structure or signal that binds to a receptor. However, to date, no specific signal has been identified within the sequence or conformation of signal peptides. Here, the effects of various structural alterations on signal peptide function are examined. More than 50 altered signal peptides were generated by incorporation of amino acid analogs during protein synthesis. Most of these structural changes of signal peptides did not block their function; only five cases were noted in which segregation of secretory proteins was impaired. The general conclusion from this study and from other available data is that signal peptides express only a low degree of structural specificity that is incompatible with their serving as ligands for a high-affinity receptor, as suggested in the Signal Hypothesis. A new model, termed the "Allosteric Model," is presented to explain how signal peptides are recognized despite their considerable structural diversity.

Introduction

All proteins, except for a few proteins made in mitochondria or chloroplasts, are synthesized by ribosomes in the cytoplasmic compartments. Maintenance of cellular homeostasis consequently necessitates efficient mechanisms for sorting and dispatching proteins to their functional loci. Secretory proteins represent the class of proteins that has been investigated most extensively with respect to segregation from the cytoplasm, in this case by the endoplasmic reticulum

(ER) (1–5). The segregation of secretory proteins is additionally significant because it appears to serve as an accurate paradigm for other classes of proteins such as lysosomal and membrane proteins, which are also synthesized by ribosomes bound to the ER (1–5). The critical issue regarding the specificity of segregation of secretory proteins is how ribosomes synthesizing this class of proteins are selected for attachment to the ER. Milstein and coworkers (6) first disclosed the key observation that secretory proteins have short-lived amino terminal extensions of their peptide chains that distinguish these proteins from cytoplasmic proteins. A variety of names have been applied to these peptide extensions, including terms such as prepeptide, leader peptide, extra piece, and signal peptide. The latter term has received greatest currency and is used here.

A number of models have been proposed to explain how signal peptides mediate selection of secretory proteins by the ER. Currently, the most widely accepted hypothesis on the function of signal peptides, presented by Blobel and Dobberstein in 1975 (7), postulates that these peptides bind directly to a highly specific receptor. One difficulty with this model, however, has been the inability to define within signal peptides any unique structural determinant to serve as a specific ligand for a receptor, *i.e.*, the signal has not been identified. At the time the Signal Hypothesis was proposed, the expectation was that the signal would consist of strong sequence homology among signal peptides (8). When more sequence data were accrued, and no signal was discerned in the primary structure of signal peptides, it was presumed that the signal must consist of a unique conformation or secondary structure. However, predictive analysis of the secondary structure of signal peptides does not reveal any common pattern among all signal peptides (9–11).

Altogether, signal peptides exhibit a confounding degree of structural diversity for peptides that are believed to serve a common function. In addition to having a sequence heterogeneity, signal peptides vary in length (15 to 30 amino acid residues), in charge, and in amino acid composition. The amino terminal ends of most signal peptides have one or more basic amino acid residues, but there are at least eight known exceptions (12) to this rule, including two proteins that have acidic rather than basic amino acids in this region. Sequence analysis of a large number of signal peptides reveals little homology among signal peptides of unrelated proteins. On the contrary, computer analysis suggests that amino acid sequences of the central segments of signal peptides result from random sequence divergence, except that there is a bias against charged amino acid residues and a preference for leucine residues (13, 14). The main selection pressure appears to be maintenance of the hydrophobicity of this segment (13, 14), and this is an important consideration in the new model proposed here. Only two features have been noted that are universal among known signal peptides: 1) They have a central core of eight or more nonpolar amino acid residues; and 2) their carboxyl termini consist of small nonpolar residues—glycine, alanine, serine, cysteine, and threonine (1–5, 12). These two characteristics appear to be too nonspecific to constitute the proposed signal.

Despite substantial sequence heterogeneity among peptides and difficulty in defining a signal, some structural constraints are clearly imposed on signal peptides, because modification of their sequences by incorporation of amino acid analogs (15) or by mutations (16, 17) can prevent segregation of secretory proteins. The present study examined whether incorporation of a variety of amino acid analogs into signal peptides affected the segregation of several secretory proteins. More than 50 different, altered signal peptides were generated with the objective of clarifying the structure-activity relationship of signal peptides. Defining this relationship turned out to be difficult simply because most changes had no observed effect on function. The overall conclusion from the data is that very little structural specificity is expressed by signal peptides, which is the same conclusion suggested by the variability of their naturally occurring amino acid sequences. The best candidate for a signal appears to be the central nonpolar segment of signal peptides. If changes are introduced that substantially decrease the hydrophobicity of this segment, segregation of secretory proteins is inhibited. Thus, there does not appear to be complete absence of a signal, but it takes the unexpected form of a general structural characteristic of signal peptides—a central segment of nonpolar amino acids. This signal is too nonspecific to be a ligand for a high-affinity receptor as proposed by the Signal Hypothesis. An alternate hypothesis termed the "Allosteric Model" is presented to explain the seeming paradox of how a relatively nonspecific series of nonpolar amino acids can mediate highly selective segregation of secretory proteins.

Effects of Amino Acid Analogs on the Segregation of Secretory Proteins

A large variety of structural modifications was introduced into signal peptides by using numerous amino acid analogs to modify several secretory proteins that had signal peptides with dissimilar sequences. Methods used for these studies are detailed in previous reports (15, 18, 19). The effects of incorporating these various amino acid analogs in the processing of the secretory proteins was examined in the Krebs ascites cell-free system and in the intact cells of isolated rat pituitaries. Effects of the analogs were assessed by electrophoresis of newly synthesized protein to determine whether there was any inhibition of the cleavage of signal peptides. At least partial transport of a secretory protein into the ER is required to allow excision of its signal peptide (7). In many cases, the transport of secretory proteins into the ER was also checked by examining the susceptibility to proteases added after completion of protein synthesis. Proteins that had been transported successfully into microsomal vesicles were inaccessible to exogenous proteases.

Using these assays for the transport of proteins into the ER, the general finding was that the structure of signal peptides could be altered at a large number of sites

Table 2.1. Amino Acid Analogs With No Effect on the Transport of Particular Secretory Proteins[a]

	Cell-Free System	
Protein	Analog[a]	
α-subunit of hCG	Selenocysteine	O-Methylthreonine
	3,4-Dehydroproline	Threo-β-fluoroasparagine (57)
	Cis-4-hydroxyproline	Threo-β-hydroxyleucine (15)
	Threo-β-hydroxynorvaline (56)	
	4-Fluorohistine	α-Amino-β-chlorobutyric acid
	(O-methylthreonine + threo-β-hydroxyleucine	
	(α-Amino-β-chlorobutyric acid + threo-β-hydroxyleucine)	
	(Azetidine-2-carboxylic acid + 4-fluorophenylalanine + canavanine)	
β-subunit of hCG	Threo-β-fluoroasparagine (57)	
Human placental lactogen	Canavanine	Azetidine-2-carboxylic acid
	β-2-Thienylalanine	4-Fluorophenylalanine
	3,4-Dehydroproline	Threo-β-phenylserine
	Cis-4-hydroxyproline	γ-Fluoroisoleucine
	4-Thiaisoleucine	α-Amino-β-chlorobutyric acid
	(Threo-β-phenylserine + azetidine-2-carboxylic acid + 5,5,5,-trifluoroleucine)	
	(Canavanine + 4-fluorophenylalanine + azetidine-2-carboxylic acid)	
	(Canavanine + 4-fluorophenylalanine + azetidine-2-carboxylic acid + 5,5,5-trifluoroleucine)	
Bovine prolactin	Threo-β-hydroxynorvaline	α-Amino-β-chlorobutyric acid
Rat prolactin	Threo-β-hydroxynorvaline	
β-subunit of LH	Threo-β-hydroxynorvaline (56)	

	Pituitary Cells	
Protein	Analog	
Rat prolactin	Ethionine	Threo-β-hydroxynorvaline (58)
	Cis-4-hydroxyproline	Threo-β-phenylserine
	6-Fluorotryptophan	α-Amino-β-chlorobutyric acid
	Azetidine-2-carboxylic acid	4-Fluorophenylalanine
	S-2-aminoethylcysteine	
	(Canavanine + S-2-aminoethylcysteine)	
Rat growth hormone	4-Thiaisoleucine	Threo-β-hydroxynorvaline (58)
	Cis-4-hydroxyproline	Para-fluorophenylalanine
	4-Fluorohistidine	Azetidine-2-carboxylic acid
	Ethionine	α-Amino-β-chlorobutyric acid
	Meta-tyrosine	Threo-β-phenylserine
	3-Fluorotyrosine	S-2-aminoethylcysteine

Table 2.1. (Continued)

		Pituitary Cells
Protein	Analog	
Rat growth hormone	6-Fluorotryptophan (Canavanine + S-2-aminoethylcysteine)	
Corticotropin precursor	Canavanine (59)	

References are indicated in parentheses for results that have been reported previously.
[a] There is evidence for incorporation of each of these analogs by the ascites cell system (18), except for 5,5,5-trifluoroleucine.

by incorporation of amino acid analogs without affecting the transport of secretory proteins. No effect on transport of several proteins was observed with analogs of proline, threonine, phenylalanine, arginine, lysine, isoleucine, valine, asparagine, and histidine, even when two or more analogs were used simultaneously. Table 2.1 lists the combinations of amino acid analogs and the proteins that were modified without any effect on their transport into microsomal vesicles. From the results in Table 2.1, it is apparent that signal peptides can be modified at multiple sites without appreciably affecting the transport of secretory proteins into microsomal vesicles in a cell-free system or into the ER in cells. Amino acid analogs could be introduced at any region within signal peptides without interfering with their function—an unexpected finding if highly specific structure is required for binding to a receptor. Careful consideration was attended to the possibility that the analogs may not have been incorporated efficiently into proteins in these systems, although incorporation of these analogs has been documented in other systems (18). This possibility was excluded by several results. In the same systems, other types of protein processing, such as glycosylation and proteolysis (reviewed in [20]), could be inhibited by some of these analogs. The electrophoretic mobility of proteins synthesized in the presence of amino acid analogs was slightly altered, suggesting a subtle change in their structure, and each amino acid analog blocked incorporation of the cognate amino acid (18). Validity of the negative results is also supported by positive controls—cases in which transport of a secretory protein is blocked by incorporation of amino acid analogs (15).

Among the many analogs tested, only the leucine analog β-hydroxyleucine (Hle) (15) affects the transport of secretory proteins into microsomes, and this analog requires substitution at multiple sites within signal peptides before it exerts any effect. Transport of proteins such as rat apolipoprotein A-I and the α-subunit of human chorionic gonadotropin (hCG), with up to four leucine residents in their signal peptides, is not inhibited (15, 21). Transport of proteins with more leucines, such as rat prolactin and human placental lactogen (hPL), is partially inhibited (15). Finally, proteins with many leucines in their signal peptides, such as the β-subunit of hCG with seven leucines and bovine prolactin with eight

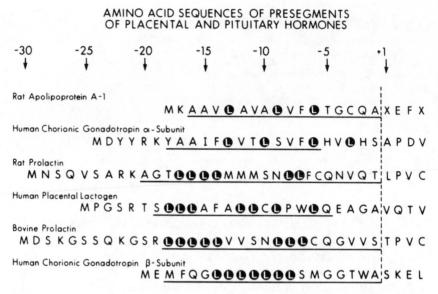

Fig. 2.1. Amino acid sequences of several signal peptides modified by incorporation of amino acid analogs. Amino acid residues are numbered from the site of cleavage by signal peptidase. Hydrophobic segments of the signal peptides are underlined, and leucine residues are circles.

leucines, have their transport completely inhibited (15, 22). (Sequences of these signal peptides and the distribution of leucine residues are indicated in Figure 2.1.) Also, it has been noted that processing of the precursor of procine relaxin, which has seven leucines dispersed through its signal peptide (23), is partially inhibited by Hle (24). Apparently, inhibition of the function of signal peptides requires a gross alteration in their structure and properties. Hle is unusual among amino acid analogs in that its physical properties differ considerably from those of the amino acid leucine, for which it substitutes. Addition of a hydroxyl group to the β position of leucine substantially decreases the hydrophobicity of this side chain. Consequently, multiple substitutions of Hle for leucine are expected to have an extreme effect on the structure and hydrophobicity of signal peptides. Clustering of leucine residues in the signal peptide of bovine prolactin and the β-subunit of hCG also would accentuate the local effect of Hle on signal peptides. Inhibition of the segregation of bovine prolactin and the β-subunit of hCG by Hle suggests the additional conclusion that one region within signal peptides that is essential for their function is the central nonpolar segment. The signal peptides of these two proteins have leucine residues only within this segment. Consequently, only this portion of signal peptides is altered by incorporation of Hle into these secretory products, while their transport into the ER is prevented.

As shown above, the inhibitory effect of Hle on the processing of secretory proteins correlates directly with the number of leucine residues in the signal peptide of each protein. A second approach to examining this correlation is to assess the effect of Hle at different concentrations that allow Hle to substitute for varying proportions of leucine residues. Previous experiments cited here used high concentrations of amino acid analogs to achieve maximal substitution of analogs for natural amino acids. The experiment presented in Figure 2.2 examines the effect of different submaximal concentrations of Hle on processing of hPL and bovine prolactin. These proteins were synthesized in a cell-free system and analyzed by electrophoresis. The top panels of Figure 2.2 show the total products of translation of human placental mRNA and bovine pituitary mRNA that, respectively, direct the synthesis of hPL and prolactin as the major product (15). Hle at the indicated concentration and microsomal membranes were included in each reaction. The lower panels show the products that are resistant to added protease due to sequestration within vesicles. Progressively greater inhibition of transport

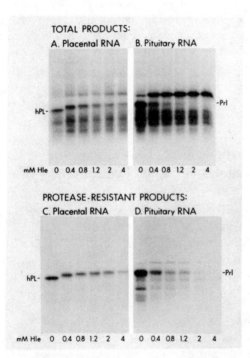

Fig. 2.2. Concentration dependence of the effect of Hle on the processing of bovine pre-prolactin and human pre-placental lactogen. Human placental RNA and steer pituitary RNA were translated by the ascites cell-free system containing microsomal membranes. Reactions contained [^{35}S]methionine and the indicated concentrations of Hle. Half of each reaction was precipitated with trichloroacetic acid after translation was completed. The other half was incubated 30 min. with a mixture of trypsin and chymotrypsin, each 20 μg/ml, before precipitation with trichloroacetic acid. Aliquots of each sample were analyzed by polyacrylamide gel electrophoresis.

and processing of the preforms of hPL and bovine prolactin occurred as the concentration of Hle increased. Again, it appears that there was a direct correlation between the amount of Hle incorporated and the inhibition of processing. Concentration-dependent substitution of Hle for an increasing number of leucines is apparent from the stepwise decrease in the electrophoretic mobility of hPL as the concentration of Hle increased. Also, estimates of the percent substitution of Hle for leucine at the different Hle concentrations were made based on the competitive inhibition of [^{3}H]-leucine incorporation by Hle. Incorporation of label was inhibited 40% at 0.4 mM Hle, 60% at 0.8 mM, 75% at 1.2 mM, 85% at 2 mM, and 95% at 4 mM. These figures suggest that substitution of Hle for leucine is required at several sites within these proteins before their processing is inhibited.

From these results, the effect of Hle appears to have a threshold. Up to about four substitutions of Hle for leucine within a signal peptide are permitted without inhibiting the processing of a secretory protein. Increasing substitution of Hle beyond this number progressively inhibits the processing of a secretory protein. This interpretation argues against the possibility that Hle has an effect by substituting at a single critical site within signal peptides. Instead, a cumulative effect of many substitutions is required. It is interesting to note that the functional transformation by Hle of secretory proteins into cytoplasmic proteins occurs as a gradual continuum; signal peptides of some proteins such as hPL may be partially inactivated, so that half of the product is secretory and half is cytoplasmic.

Comparison of the Effects of Mutations and Amino Acid Analogs on Signal Peptides

In bacteria, it has been possible to apply genetic methods, rather than amino acid analogs, to modify signal peptides. However, there are similarities in the conclusions reached with either approach. Point mutations in bacterial signal peptides, which prevent the transport of their parent secretory proteins out of bacteria (16, 17), consist primarily of substitution of a charged for a noncharged amino acid. This is the most drastic structural change possible with a point mutation; and it suggests, like the results with Hle, that inactivation of the signal function of signal peptides requires a major alteration in their structure. Probably, many other mutations occured within signal peptides in which there was no change in charge, but these mutations did not affect the processing of a secretory protein and, thus, were not detected by the selection techniques. The point mutations that blocked protein translocation occurred mainly within the central hydrophobic segment of signal peptides, again suggesting that the signal lies within this region, but no single critical site was identified; the point mutations could occur at several points within signal peptides. Recent application of site-directed mutagenesis to the amino terminal end of a bacterial signal peptide has shown that changes in the charge of this region slow, but do not block, export of the protein, indicating that this region is less critical (25). Similar results are reported by Hall *et al.* (26), who conclude that modification of this region of signal peptides affects a different function than does modification of hydrophobic segments of signal peptides.

One notable difference between the effect of Hle and the point mutations is that Hle requires substitution at multiple sites. Addition of hydroxyl groups alters the properties of amino acids less dramatically than addition of a charged group. (Several times as much energy is required to transfer a charge group into a nonpolar environment as to transfer a hydroxyl group [13, 14].) Thus, multiple substitutions of Hle are required to yield the same effect as addition of a single charged amino acid.

Evolution of protein sequences provides experiments of nature on permissible mutations of signal peptides. Fairly weak sequence conservation of signal peptides is evidenced not only by the previously noted sequence heterogeneity, but also by the rate of sequence change in related signal peptides. For example, comparison of the two genes for rat preproinsulin (27) finds a rate of sequence change in signal peptides greater than that in the C peptide of proinsulin, which is believed to serve only as a spacer for the assembly of insulin chains. In contrast, with the substantial sequence divergence of signal peptides, their function is surprisingly conserved through evolution. Microsomes prepared from one tissue source will process secretory proteins from a large variety of tissues and species, even from plants (2, 4, 5). At an even greater extreme, when the gene for rat preproinsulin is inserted into bacteria, proinsulin is secreted (28). In this case, *Escherichia coli* must recognize the signal within a rat signal peptide. The converse has also been observed: that the bacterial protein β-lactamase is transported by eukaryotic systems (29, 30). The explanation offered here for this great conservation of function is that a general property of signal peptides, such as their hydrophobicity, is recognized rather than a highly specific conformation. It seems less likely that a receptor could be conserved as stringently across evolution so as to universally bind specific ligands produced by phylogenetically distant organisms.

Mechanism of the Effect of Hle on the Transport of Secretory Proteins

Recently, the laboratories of Blobel and Dobberstein have made great progress in characterizing components of the ER that are required for the transport of secretory proteins into the ER. To date, two major components have been characterized: an integral membrane protein that has been termed the docking protein (31–34), and a ribonucleoprotein complex called the signal recognition particle (SRP), which is readily released from the ER (35–39). During the characterization of SRP, Walter and Blobel clarified the mechanism by which Hle inhibits the transport of bovine prolactin into microsomes (35). Usually, SRP binds to ribosomes synthesizing secretory proteins such as prolactin; and SRP mediates attachment of the ribosomes to the ER via the docking protein. Incorporation of Hle into nascent prolactin inhibits the association of SRP with ribosomes, thereby preventing the attachment of ribosomes to the ER. This result clearly shows that the association of SRP with ribosomes depends on the structure of nascent peptides undering synthesis. Walter and Blobel propose that Hle modifies the structure of the signal peptide for prolactin and thereby abolishes its binding to SRP, considering SRP to be a signal receptor. However, as noted previously here, indirect evidence suggests that signal peptides lack a suitable ligand for binding to a receptor. Part of the solution to this dilemma may be that SRP has a primary role as a ribosome receptor, rather than as a signal receptor. Several results by Walter and Blobel suggest that SRP acts, at least in part, as a ribosome receptor. SRP binds with appreciable affinity to ribosomes not engaged in protein synthesis (dissociation constants of about 6×10^{-5} vs. 8×10^{-9} for polysomes synthesizing prolactin, and ribosome binding of SRP, as well as its function, were abolished by treatment with N-ethylmaleimide and with heat [35]). When SRP associates with ribosomes synthesizing prolactin, translation is arrested until the ribobome-SRP complex binds to microsomes (37). An effect on translation such as this suggests that SRP binds directly to ribosomes rather than to signal peptides. Nascent peptide chains exist from the large ribosomal subunit distant from the site of peptide bond synthesis (39, 40); and, thus, it is improbable that binding of SRP to signal peptides would sterically hinder peptide bond synthesis. On the other hand, many agents that bind to ribosomes are known to inhibit protein synthesis. Two attractive possibilities for the binding of SRP to ribosomes are suggested by recent identification of a small RNA molecule in SRP (38). The small RNA could form base pairs with ribosomal RNA or could interact with ribosomal proteins, since many of these proteins are functionally adapted to bind RNA.

An Allosteric Model for Recognition of Signal Peptides

The major limitation to assigning SRP a primary role as a ribosome receptor is to account for its specific high-affinity binding of ribosomes synthesizing secretory proteins. This limitation can be overcome by postulating that signal peptides affect the conformation of ribosomes, and thereby increase the affinity of ribosomes for SRP. Designation of this model of signal recognition as the Allosteric Model derives from its proposal that signal peptides indirectly enhance the binding of ribosomes to SRP, rather than binding with high-affinity directly to SRP. Comparison of the Allosteric Model with the Signal Hypothesis is presented schematically in Figure 2.3.

The Signal Hypothesis (shown along the left side of Figure 2.3) proposes that ribosomes are free until signal peptides are synthesized. Then SRP binds tightly to signal peptides, inhibiting protein synthesis and mediating attachment of ribosomes to the ER via the docking protein. Attachment of ribosomes to the ER is solidified by binding to a separate ribosome receptor. Protein synthesis resumes

MODELS FOR SIGNAL RECOGNITION

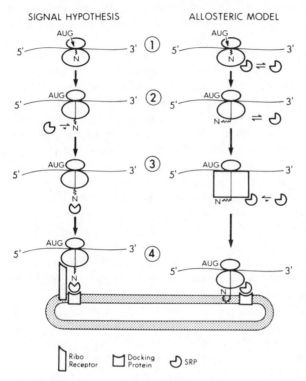

Fig. 2.3. Schematic comparison of the Signal Hypothesis and the Allosteric Model. Details of these models are described in the text.

after ribosomes are bound to the ER, and secretory proteins are inserted into the ER through a proteinaceous pore.

The Allosteric Model differs significantly from the Signal Hypothesis in a number of respects. In the Allosteric Model, ribosomes bind weakly to SRP even before signal peptides are synthesized; *i.e.*, there is a dynamic equilibrium as ribosomes rapidly bind and release. As signal peptides are synthesized and reach the surface of ribosomes, they resist extension into the aqueous cytoplasm due to their hydrophobicity; instead, they interact with a hydrophobic site on ribosomes, resulting in a conformational change of ribosomes to a state in which protein synthesis is arrested and SRP is bound much more tightly. SRP then implements attachment of ribosomes to the ER via the docking protein, as in the Signal Hypothesis, and there may be no need for a separate ribosome receptor because SRP serves as a ribosome receptor. After ribosomes bind to the ER, signal peptides release from their site on ribosomes and insert into the membrane of the ER, probably into the lipid bilayer, by virtue of their hydrophobicity. (Their affinity for the bilayer is similar to or greater than their affinity for the site on ribosomes.) This allows ribosomes to return to their low-affinity conformation (with respect to SRP), with active protein synthesis and eventual release of ribosomes from the ER after synthesis of their nascent peptide chains is completed.

In addition to explaining why SRP binds ribosomes, the Allosteric Model of signal peptide recognition offers explanations for several findings that have been difficult to reconcile with the Signal Hypothesis. The foremost advantage of the Allosteric Model is that it explains how signal peptides are recognized despite their structural diversity. This model does not require signal peptides to have a highly specific conformation for recognition by a receptor, because signal peptides are oriented and brought to their site of action on the ribosome in high concentration by the attachment of nascent peptide chains to ribosomes. Consequently, the affinity of signal peptides for their binding site on ribosomes can be quite low and consist of a rather nonspecific interaction. Binding of the central nonpolar segment of signal peptides with a hydrophobic site on ribosomes would be adequate. This could explain the enigmatic nature of the signal within signal peptides and account for the inability to find a specific sequence or conformation, as predicted by the Signal Hypothesis. The Allosteric Model tolerates minor modifications in the structure of signal peptides, as with the incorporation of most amino acid analogs and with many mutations, except those that drastically change the hydrophobicity of the central portion of signal peptides. Probably, there are only general conformational requirements for the signal peptide to permit appropriate contact with the central hydrophobic segment. Evidence for some conformational restriction is provided by two cases in which signal peptides are inactivated by mutations that introduce helix breakers (41).

Other advantages offered by the Allosteric Model in comparison with the original Signal Hypothesis are:

1. Low-affinity binding of signal peptides in the Allosteric Model permits dissociation of these peptides from their binding site on ribosomes, so that signal peptides can be transported at least partially into the ER. Entrance of signal peptides into the ER is necessary before these peptides can be cleaved from their parent polypeptides (7). Currently, the Signal Hypothesis does not explain how signal peptides dissociate from their receptor.

2. The Allosteric Model provides considerable signal amplification. The SRP-binding site can consist of a substantial part of a ribosome that has a molecular weight of about 4 million daltons (42), while signal peptides present a relatively minuscule target of as few as 15 amino acid residues. SRP would need to contact ribosomes hundreds or even thousands of times before fortuitously encountering the signal peptide, as required in the Signal Hypothesis. To make direct contact of SRP with signal peptides even more improbable, their marked hydrophobicity should diminish their extension into the aqueous cytoplasm; instead, the critical hydrophobic segment of signal peptides should preferentially remain against the surface or within ribosomes. In the Allosteric Model, SRP would require many fewer contacts with ribosomes before encountering its relatively large binding

site; and in this model, SRP needs to recognize only a single structure, rather than the multitude of different sequences occurring in signal peptides.

3. The Allosteric Model explains how signal peptides can be recognized across a broad evolutionary span—how a bacterial cell can recognize eukaryotic signal peptides, as in the case of secretion of proinsulin by *E. coli* containing a cloned insulin gene (28). Understanding this is increasingly important as more eukaryotic genes are cloned and expressed in bacteria. All that is needed is for the hydrophobic segment of signal peptides to interact with a particular site on bacterial ribosomes. There is no need for tight binding of eukaryotic signal peptides to a receptor.

4. Since SRP is a low-affinity ribosome receptor in the absence of signal peptides, proteolysis of signal peptides or their transport into the ER provide mechanisms for the eventual release of ribosomes from the ER.

Evidence for the Signal Hypothesis

Most results that are cited as evidence for the Signal Hypothesis can be explained just as well in the context of the Allosteric Model. An example is inhibition of cell-free processing of secretory proteins by a chemically synthesized signal peptide (43) and by a tryptic fragment of ovalbumin (44). In these experiments, the peptides were added to cell-free systems at high concentrations (several orders of magnitude greater than the concentration of signal peptides synthesized in the cell-free systems). At these concentrations, the amount of added peptide at the surface of ribosomes will approach the local concentration of signal peptides on nascent peptide chains, and any hydrophobic sites on ribosomes can be saturated by the added peptides. This may inhibit processing of secretory proteins by two mechanisms—first by blocking the interaction of signal peptides with their site on ribosomes, and second by tending to put all ribosomes into a high-affinity state for binding of SRP, so that a limiting amount of SRP is bound unproductively. A prediction from the Allosteric Model, but not the Signal Hypothesis, is that the inhibitory peptide can be a hydrophobic peptide other than a signal peptide, because little structural specificity is required to bind to the site on ribosomes. This prediction is supported by the recent finding that the inhibitory peptide from ovalbumin is not the signal peptide for this protein (45, 46). Also, this author has observed that processing of secretory proteins is inhibited potently *in vitro* by the polypeptide antibiotic gramicidin S (22), which has no functional relationship to signal peptides. A major problem with this result, and with the effects of peptides from preproparathyroid hormone and ovalbumin, is to decide whether inhibition of processing of secretory proteins results from specific interaction with the processing apparatus or from nonspecific interaction with membranes. Agents such as detergents and calcium (reviewed in [5]) are also capable of inhibiting processing by microsomes. Examination of the effects of peptides on the binding of SRP would be helpful in determining whether peptides, in fact, have a specific effect.

Testing the Allosteric Model

The Allosteric Model is a logical explanation for available data on the recognition of signal peptides and for previous inability to identify a signal. However, equally important, the Allosteric Model is useful as a working model in directing future investigations. It suggests that further conformational analysis of signal peptides will have limited utility. Instead, this model points future studies toward investigation of the role of ribosomes in the attachment of nascent secretory proteins to the ER via SRP. The interaction of SRP with ribosomes can be explored by cross-linking and high-resolution electron microscopy; and possible base pairing between the RNA in SRP and ribosomal RNA can be explored by comparison of their base sequences. In addition, the Allosteric Model predicts that modifying or blocking certain sites on ribosomes will prevent the attachment of secretory proteins to the ER. This prediction can be tested by a large variety of approaches; and some data supporting the Allosteric Model are already available from studies that have used various techniques to modify ribosomes.

Possibly the easiest agents to use to specifically modify ribosomes are antibiotics. Interestingly, many antibiotics differentially inhibit the translation of free and membrane-bound ribosomes (47–49). There are a number of possible explanations for this phenomenon; one tantalizing possibility is that in some cases, antibiotics affect the attachment of ribosomes to the ER, and thereby change the overall ratio of free and bound ribosomes as well as the ratio of activity by the two classes of ribosomes. Sabatini's laboratory determined in 1974 that the antibiotic aurin tricarboxylic acid inhibited the binding of ribosomes to microsomes *in vitro* (50). In the following year, Blobel and Dobberstein noted that this antibiotic inhibited the cell-free processing of a secretory protein (7). A more detailed analysis has been performed for the antibiotic spectinomycin, using mutants of *E. coli* that are resistant to the drug due to alteration of ribosomal protein (51–53).

Genetic modification of ribosomes and membrane components offers one of the most flexible and powerful approaches for analyzing the role of different cellular constituents in the processing of secretory proteins. This approach is just becoming possible with higher eukaryotes. Previously, it has been applied with great success in yeast to study steps further along the secretory pathway (reviewed in [54]), and many elegant studies have been performed in bacteria. One of these genetic studies in *E. coli* by Emr *et al.* (55) affords the most direct evidence for the Allosteric Model to date. This group isolated mutations that would restore the export of lambda bacteriophage receptor protein, which had its export blocked due to mutations in its signal peptide. The suppressor mutations mapped to three different loci on the bacterial chromosome. One of the loci, the most frequently occurring, was mapped precisely. This locus was in the major ribosomal gene cluster very near the locus for spectinomycin resistance, indicating that the mutation altered a ribosomal protein. From the effects of these suppressor mutations, the authors concluded that, "the cellular component altered by the prl [protein localization] suppressor mutation may interact directly with the λ receptor signal sequence." In other words, they conclude that a ribosomal protein directly interacts with a signal peptide. Bacteria containing the suppressor mutations could export λ receptor with a variety of mutations in its signal peptide, even mutations that delete a large portion of its signal peptide, including the central hydrophobic segment. The Signal Hypothesis cannot account for this secretion of a protein that has lost its signal. In the Allosteric Model, this is possible supposing that the suppressor mutations alter ribosomes so that they bind membranes with high affinity, even when the ribosomes do not have an attached signal peptide. In this circumstance, ribosomes synthesizing any nascent protein could be bound to be cytoplasmic membrane, and some of the specificity of protein secretion would be lost. Small amounts of many proteins, even cytoplasmic proteins, might be secreted.

Acknowledgments. The experimental data presented here comprised part of the author's Ph.D. thesis in the laboratory of Dr. Irving Boime (supported by grant HD-13481 from the National Institutes of Health and by funds to the author from the NIH Medical Scientist Training Program, grant T-32 GM-07022). I thank Dr. Boime and Dr. Arnold Strauss for review of this manuscript. Amino acid analogs were generously provided by Drs. Robert Abeles, Herman Gershon, Robert Handschumacher, Kenneth Klein, Theodore Otani, Uppendra Pandit, Marco Rabinowitz, and Janice Sufrin. Vickey Farrue assisted with preparation of the manuscript.

References

1. Blobel, G. (1982) Harvey Lect. *76,* 125–147.
2. Kreil, G. (1981) Ann. Rev. Biochem. *50,* 317–348.
3. Meyer, D. I. (1982) Trends Biochem. Sci. *7,* 320–321.
4. Sabatini, D. D., Kreibich, G., Morimoto, T. & Adesnik, M. (1982) J. Cell Biol. *92,* 1–22.
5. Strauss, A. W. & Boime, I. (1982) CRC Crit. Rev. Biochem. *12,* 205–236.
6. Milstein, C., Brownlee, G. G., Harrison, T. M. & Mathews, M. B. (1972) Nature New Biol. *239,* 117–120.
7. Blobel, G. & Dobberstein, B. (1975) J. Cell Biol. *67,* 835–851.
8. Devillers-Thiery, A., Kindt, T., Schecle, G. & Blobel, G. (1975) Proc. Natl. Acad. Sci. USA *72,* 5016–5020.
9. Austen, B. M. (1979) FEBS Lett. *103,* 308–313.
10. Garnier, J., Gaye, P., Mercier, J. C. & Robson, B. (1980) Biochemie *62,* 231–239.
11. Chan, S. J., Patzelt, C., Duguid, J. R., Quinn, P., Labrecque, A., Noyes, B., Keim, P., Heinrikson, R. L. & Steiner, D. F. (1979) In *From Gene to Protein: Information Transfer in Normal and Abnormal Cells* (pp. 361–378), Russell, T. R., Brew, K., Faber, H. & Schultz, J., eds. Academic Press, New York.

12. Perlman, D. & Halvorsen, H. O. (1983) J. Mol. Biol. *167*, 391–410.
13. Von Heijne, G. (1980) Eur. J. Biochem. *103*, 431–438.
14. Von Heijne, G. (1981) Eur. J. Biochem. *116*, 419–422.
15. Hortin, G. & Boime, I. (1980) Proc. Natl. Acad. Sci. USA *77*, 1356–1360.
16. Emr, S. D., Hedgepeth, J., Clement, J. M. & Silhavy, T. J. (1980) Nature *285*, 82–85.
17. Bedouelle, H., Bassford, P. J., Jr., Fowler, A. V., Zabin, I., Beckwith, J. & Hofnung, M. (1980) Nature *285*, 78–81.
18. Hortin, G. & Boime, I. (1983) Meth. Enzymol. *96*, 777–784.
19. Hortin, G. & Boime, I. (1981) Cell *24*, 453–461.
20. Hortin, G. & Boime, I. (1983) Trends Biochem. Sci. *8*, 320–323.
21. Gordon, J. I., Smith, D. P., Andy, R., Alpers, D. H., Schonfeld, G. & Strauss, A. W. (1982) J. Biol. Chem. *257*, 971–978.
22. Hortin, G. (1983) Ph.D. Thesis. Washington University, St. Louis, MO.
23. Haley, J., Hudson, P., Scanlon, D., John, M., Cronk, M., Shine, J., Tregear, G. & Niall, H. (1982) DNA *1*, 155–162.
24. Gast, M. J. (1983) J. Biol. Chem. *258*, 9001–9004.
25. Vlasuk, G. P., Inouye, S., Ito, H., Itakura, K. & Inouye, M. (1983) J. Biol. Chem. *258*, 7141–7148.
26. Hall, M. N., Gabay, J. & Schwartz, M. (1983) EMBO J. *2*, 15–19.
27. Lomedico, P., Rosenthal, N., Efstsratidadis, A., Gilbert, W., Kolodner, R. & Tizard, R. (1979) Cell *18*, 545–558.
28. Talmadge, K., Stahl, S. & Gilbert W. (1980) Proc. Natl. Acad. Sci. USA *77*, 3369–3373.
29. Muller, M., Ibrahimi, I., Chang, C. N., Walter, P. & Blobel, G. (1982) J. Biol. Chem. *257*, 11860–11863.
30. Roggenkamp, R., Kusterman-Kuhn, B. & Hollenberg, C. P. (1981) Proc. Natl. Acad. Sci. USA *78*, 4466–4470.
31. Meyer, D. I., Krause, E. & Dobberstein, B. (1982) Nature *297*, 647–650.
32. Meyer, D. I., Louvard, D. & Dobberstein, B. (1982) J. Cell. Biol. *92*, 579–583.
33. Gilmore, R., Blobel, G. & Walter, P. (1982) J. Cell. Biol. *95*, 463–469.
34. Gilmore, R., Walter, P. & Blobel, G. (1982) J. Cell. Biol. *95*, 470–477.
35. Walter, P., Ibrahimi, I. & Blobel, G. (1981) J. Cell. Biol. *91*, 545–550.
36. Walter, P. & Blobel, G. (1981) J. Cell. Biol. *91*, 551–556.
37. Walter, P. & Blobel, G. (1981) J. Cell. Biol. *91*, 557–561.
38. Walter, P. & Blobel, G. (1982) Nature *299*, 691–697.
39. Blobel, G. & Sabatini, D. (1970) J. Cell. Biol. *45*, 130–145.
40. Malkin, L. I. & Rich, A. (1967) J. Mol. Biol. *26*, 329–346.
41. Emr, S. D. & Silhavy, T. J. (1983) Proc. Natl. Acad. Sci. USA *80*, 4599–4603.
42. Wool, I. G. (1979) Ann. Rev. Biochem. *48*, 719–754.
43. Majzoub, J. A., Rosenblatt, M., Fennick, B., Maunus, R., Kronenberg, H. M., Potts, J. T., Jr. & Habener, J. F. (1980) J. Biol. Chem. *265*, 11478–11483.
44. Lingappa, V. R., Lingappa, J. R. & Blobel, G. (1979) Nature *281*, 117–121.
45. Palmiter, R. D., Gagnon, J. & Walsh, K. A. (1978) Proc. Natl. Acad. Sci. USA *75*, 94–98.
46. Braell, W. A. & Lodish, H. F. (1982) J. Biol. Chem. *257*, 4578–4582.
47. Glazer, R. I. & Sartorelli, A. C. (1972) Biochem. Biophys. Res. Commun. *46*, 1418–1425.
48. Hirashima, A., Childs, G. & Inouye, M. (1973) J. Mol. Biol. *79*, 373–389.
49. Halegoua, S., Hirashima, A. & Inouye, M. (1976) J. Bacteriol. *126*, 183–191.

50. Borgese, N., Mok, W., Kreibich, G. & Sabatini, D. D. (1974) J. Mol. Biol. *88*, 559–580.
51. Dombou, M., Mizuno, T. & Mizushima, S. (1977) Mol. Gen. Genet. *155*, 53–60.
52. Mizuno, T., Yamada, H., Yamagata, H. & Mizushima, S. (1977) J. Bacteriol. *125*, 524–530.
53. Mizuno, T., Yamagata, H. & Mizushima, S. (1977) J. Bacteriol. *129*, 326–332.
54. Schekman, R. (1982) Trends Biochem. Sci. *7*, 243–246.
55. Emr, S. D., Hanley-Way, S. & Silhavy, T. J. (1981) Cell *23*, 79–88.
56. Hortin, G. & Boime, I. (1980) J. Biol. Chem. *255*, 8007–8010.
57. Hortin, G., Stern, A. M., Miller, B., Abeles, R. H. & Boime, I. (1983) J. Biol. Chem. *258*, 4047–4050.
58. Hortin, G. & Boime, I. (1981) J. Biol. Chem. *256*, 1491–1494.
59. Hoshina, H., Hortin, G. & Boime, I. (1982) Science *217*, 63–64.

3

Proteolytic Processing of Signal Peptides

Henry C. Wu

It is now generally accepted that most proteins destined for export contain NH_2 terminal signal peptides that are cleaved during or immediately following their synthesis. Ever since the formulation of signal hypothesis by Blobel and Dobberstein (1), the role of signal sequence in protein export has been a topic of intensive studies, especially in recent years. The rapid pace of research in this area can be attributed to two major advances in the field: 1) the sequences of precursor proteins can be readily deduced by the cloning and DNA sequencing of genes encoding exported proteins in both eukaryotic and prokaryotic cells; and 2) molecular genetics and recombinant DNA technology are used to modify the signal sequences of exported proteins. The successful applications of both approaches plus the significant progress made by Blobel, Walter, Dobberstein, and Meyer on the reconstitution of secretory machinery in microsomal fractions (2–10) have permitted the formulation of a more detailed model regarding the role of signal sequences in protein export processes. Several questions have been posed concerning the structures and functions of signal sequences:

1. Are there common features in the primary or secondary structures of signal peptides found in a diverse group of exported proteins?
2. What are the roles of these common structural features shared by signal peptides of exported proteins in the export process?
3. What is the role of proteolytic cleavage of the signal peptide by signal peptidase in the export process?

This review will summarize our current knowledge concerning both the structure and functions of signal sequences and of signal peptidases in protein export processes.

Signal Hypothesis Revisited

Recent work of Blobel, Walter, Dobberstein, Meyer, and their coworkers (2–10) has refined the signal hypothesis that was originally formulated to account for the synthesis of secretory proteins on membrane-bound polysomes and for the presence of NH_2 terminal signal sequences in the precursor forms of secretory proteins. The current signal hypothesis model (Figure 3.1) (11, 12) postulates a specific interaction between the signal sequence of a nascent chain and the signal recognition particle (SRP) in the cytoplasm. This interaction results in a translational block in the elongation of the nascent chains of secretory proteins. In the second stage of the export process, the synthesis of exported proteins is resumed upon the binding of SRP:signal peptide complex to a specific SRP receptor protein (the docking protein) located on the cytoplasmic surface of the rough endoplasmic reticulum (ER) membrane. While these initial steps in protein secretion appear well documented with experimental evidence, the subsequent step (*i.e.*, the mechanism by which the nascent precursor protein is translocated across the membrane during or immediately after the synthesis) remains the most obscure step in the overall process of protein secretion. The final step of protein export is the removal of the NH_2 terminal signal peptide by proteolytic cleavage, which presumably takes place at the luminal side of the membrane.

An alternative hypothesis has been proposed by Wickner, who envisions the function of the signal peptide in modifying the folding pathway of the precursor protein (Figure 3.2) (13, 14). In its simplest version, protein secretion, according to the membrane trigger hypothesis, is a self-assembly process based on protein-

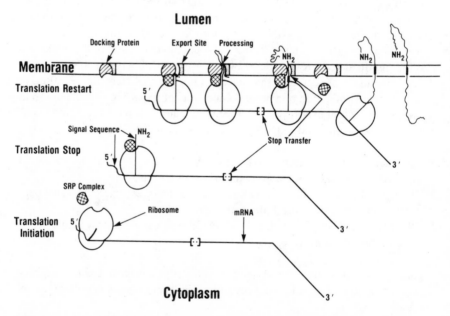

Fig. 3.1. Cotranslational protein export according to the signal hypothesis. SRP, signal recognition particle. (Taken from [12].)

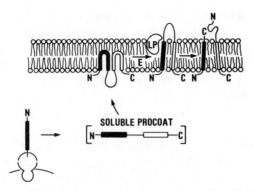

Fig. 3.2. Membrane-triggered folding hypothesis for the assembly of M13 procoat protein. Thin lines refer to polar residues, open thick lines to hydrophobic sequences, and filled thick lines to the hydrophobic part of the signal sequence. (Taken from [14].)

protein and protein-lipid interactions with little intervention of catalysis. The best example supporting this model is the assembly of M13 procoat protein into liposomes. The signal hypothesis and membrane trigger hypothesis are not mutually exclusive. Recent results from Randall's laboratory (15) can be best accounted for by a combination of these two working hypotheses; *i.e.*, the initial stage of signal recognition as envisioned by the signal hypothesis, followed by folding of nascent polypeptides on the cytoplasmic membrane that may be triggered or affected by the components or hydrophobic environments of the membrane (16). Translocation of the nascent chain across the membrane may ensue when a large domain of the nascent chain assumes a proper conformation.

It is not clear whether there is specific protein-protein interaction between the nascent chain of exported protein and membrane proteins of the putative export machinery *during* the actual translocation process across the membrane. Nor is it certain that signal sequences are involved in the recognition of precursor proteins by the export machinery within the membranes beyond the initial step of signal recognition by SRP prior to the attachment of the nascent chain to the membrane. However, it is reasonable to assume that there are special structural features contained in the signal peptides of exported proteins that are important for the correct cleavages of precursor proteins by signal peptidases. In the following sections, we shall examine the structures of signal peptides for common features that may be important for at least two distinct and specific interactions of signal peptides with other cellular components: 1) the interaction of signal sequences with SRP or putative export machinery; and 2) the interaction of signal peptides with signal peptidases.

Structures of Signal Peptides

It is obvious from inspection of known signal sequences that they vary considerably in length, in both amino acid composition and primary structure. Two approaches have been employed to define common structural features shared by seemingly unrelated signal sequences. The first approach entails the use of statistical analyses of known signal sequences from diverse sources. Certain common features have emerged from these analyses. The second approach relies on the isolation and characterization of mutants with random, selected, or predetermined alterations in the signal peptides of various exported proteins. These mutants can then be used to ascertain the functions of signal sequences in protein export.

The first approach has been employed by a number of investigators to analyze signal sequences in exported proteins in both prokaryotic and eukaryotic cells. Based on signal sequences of four precursor proteins of *Escherichia coli* (prolipoprotein, pro-β-lactamase, pro-fd-major coat protein, and pro-fd-minor coat protein), Inouye and coworkers postulated a loop model for the function of signal sequences in protein secretion (17). In this model, the structure of signal peptides was divided into three domains: 1) the hydrophilic and positively charged segment at the NH_2 terminus; 2) the hydrophobic core in the middle segment of the signal sequences; and 3) the COOH terminal segment of the signal sequence enriched in amino acids with small side chains immediately preceding the cleavage site.

Table 3.1 summarizes the signal sequences of prokaryotic precursor proteins found mostly in *E. coli* (18–42). The signal sequences of a large number of precursor proteins have many structural features in common, as first recognized by Inouye and coworkers: 1) basic amino acid residues at the NH_2 terminus of the signal peptide; 2) a sequence of 10 to 12 hydrophobic amino acids immediately following the positively charged hydrophilic segment at the NH_2 terminus; 3) alanine or glycine residue at the cleavage site; and 4) proline or glycine within the hydrophobic domain (Figure 3.3). An additional feature can be discerned: there is a nonrandom distribution of Thr/Ser residues in the signal peptides (Figure 3.3). The clustering of Thr/Ser residues around −5 from the cleavage site suggests a specific recognition of these residues by components of the export machinery or the processing enzyme. Alternatively, the presence of Thr/Ser at the COOH terminal end of the hydrophobic segment may predispose conformational transition of secondary structures from the α-helix or β sheet to the random coil/β turn, which is predominant near the cleavage site.

Predictions of secondary structures of signal sequences have been made according to the Chou-Fasman method (43), even though it is generally recognized that the Chou-Fasman empirical rule has not been validated for membrane proteins present in a hydrophobic environment. In view of the fact that there is an apparent lack of conserved primary structures of signal peptides, special emphasis has been placed on the importance of common secondary structures shared by diverse signal peptides in protein export. Two common features have emerged. First, the hydrophobic core would assume a periodic structure, either in the form of an α-helical structure or a β sheet structure. This is followed by β turn or extensive random coil region near the cleavage site. It has been further

Table 3.1. Signal Sequences of Prokaryotic-Exported Proteins

I. Gram-negative bacteria
A. Outer membrane
   proteins $\quad$ −20
   1. Lipoprotein (18) $\quad$ MetLysAlaThrLysLeuValLeuGlyAlaValIleLeuGlySerThrLeu
   +1
   LeuAlaGlyCysSer

   2. OmpA protein (19) $\quad$ −21
   MetLysLysThrAlaIleAlaIleAlaValAlaLeuAlaGlyPheAlaThrVal
   +1
   AlaGlnAlaAlaPro

   OmpA protein $\quad$ −21
   (*S. dysenteriae*) (20) $\quad$ MetLysLysThrAlaIleAlaIleThrValAlaLeuAlaGlyPheAlaThrVal
   +1
   AlaGlnAlaAlaPro

   3. Lambda receptor $\quad$ −25
   (21) $\quad$ MetMetIleThrLeuArgLysLeuProLeuAlaValAlaValAlaAlaGly
   +1
   ValMetSerAlaGlnAlaMetAlaValAsp

   4. PhoE protein (22) $\quad$ −21
   MetLysLysSerThrLeuAlaLeuValValMetGlyIleValAlaSerAla
   +1
   SerValGlnAlaAlaGlu

   5. OmpF protein (23) $\quad$ −22
   MetMetLysArgAsnIleLeuAlaValIleValProAlaLeuLeuValAla
   +1
   GlyThrAlaAsnAlaAlaGlu

   6. OmpC protein (24) $\quad$ −21
   MetLysValLysValLeuSerLeuLeuValProAlaLeuLeuValAlaGly
   +1
   AlaAlaAsnAlaAlaGlu

B. Periplasmic proteins
   7. Alkaline phosphatase $\quad$ −21
   (25) $\quad$ MetLysGlnSerThrIleAlaLeuAlaLeuLeuProLeuLeuPhe
   +1
   ThrProValThrLysAlaArgThr

   8. AmpC β-lactamase $\quad$ −19
   (26) $\quad$ MetPheLysThrThrLeuCysAlaLeuLeuIleThrAlaSerCysSerThr
   +1
   PheAlaAlaPro

   9. TEM β-lactamase $\quad$ −23
   (27) $\quad$ MetSerIleGlnHisPheArgValAlaLeuIleProPhePheAlaAlaPhe
   +1
   CysLeuProValPheAlaHisPro

   10. Arabinose-binding $\quad$ −23
   protein (28) $\quad$ MetHisLysPheThrLysAlaLeuAlaAlaIleGlyLeuAlaAlaValMet
   +1
   SerGlnSerAlaMetAlaGluAsn

Table 3.1. (Continued)

11.	Lysine-arginine-ornithine-binding protein (29)	−22 MetLysLysThrValLeuAlaLeuSerLeuLeuIleGlyLeuGlyAlaThr +1 AlaAlaSerTyrAlaAlaLeu
12.	Histidine-binding protein (29)	−22 MetLysLysLeuAlaLeuSerLeuSerLeuValLeuAlaPheSerSerAla +1 ThrAlaAlaPheAlaAlaIle
13.	Leucine-binding protein (30)	−23 MetLysAlaAsnAlaLysThrIleIleAlaGlyMetIleAlaLeuAlaIleSer +1 HisThrAlaMetAlaAspAsp
14.	Isoleucine-valine-binding protein (31)[a]	−23 MetAsnThrLysGlyLysAlaLeuLeuAlaGlyLeuIleAlaLeuAlaPhe +1 SerAsnMetAlaLeuAla
15.	Maltose-binding protein (32)	−26 MetLysIleLysThrGlyAlaArgIleLeuAlaLeuSerAlaLeuThrThr +1 MetMetPheSerAlaSerAlaLeuAlaLysIle
16.	Galactose-binding protein (28)	−23 MetAsnLysLysValLeuThrLeuSerAlaValMetAlaSerMetLeuPhe +1 GlyAlaAlaAlaHisAlaAlaAsp
17.	Ribose-binding protein (33)	−25 MetAsnMetLysLysLeuAlaThrLeuValSerAlaValAlaLeuSerAla +1 ThrValSerAlaAsnAlaMetAlaLysAsp

C. Viral proteins

18.	Major coat protein (phage M13) (34)	−23 MetLysLysSerLeuValLeuLysAlaSerValAlaValAlaThrLeuVal +1 ProMetLeuSerPheAlaAlaGlu
19.	Minor coat protein (phage M13) (35)	−18 MetLysLysLeuLeuPheAlaIleProLeuValValProPheTyrSer +1 HisSerAlaGlu

D. Enterotoxine

20.	Heat labile toxin, subunit B (36)	−21 MetAsnLysValLysCysTyrValLeuPheThrAlaLeuLeuSerSerLeu +1 TyrAlaHisGlyAlaPro
21.	Heat labile toxin, subunit A (37)	−18 MetLysAsnIleThrPheIlePhePheIleLeuLeuAlaSerProLeu +1 TyrAlaAsnGly

Table 3.1. (Continued)

II. Gram-positive bacteria		
22. α-Amylase	−31	
(*B. amyloliquefaciens*) (38)	MetIleGlnLysArgLysArgThrValSerPheArgLeuValLeuMetCys	
	+1	
	ThrLeuLeuPheValSerLeuProIleThyLysThrSerAlaValAsn	
α-Amylase	−29	
(*B. licheniformis*) (39)	MetLysGlnGlnLysArgLeuTyrAlaArgLeuLeuThrLeuLeuPhe	
	+1	
	AlaLeuIlePheLeuLeuProHisSerAlaAlaAlaAlaAlaAsn	
23. Penicillinase	−26	
(*B. licheniformis*) (40)	MetLysLeuTrpPheSerThrLeuLysLeuLysLysAlaAlaAlaValLeu	
	+1	
	LeuPheSerCysValAlaLeuAlaGlyCysAla	
24. Penicillinase	−16	
(*S. aureus*) (41)	MetLysLysLeuIlePheLeuIleValIleAlaLeuValLeuSer	
	+1	
	AlaCysAsn	
25. Penicillinase		
(*B. cereus*) (42)	(Met)MetIleLeuLysAsnLysArgMetLeuLysIleGlyIle	
	CysValGlyIleLeuGlyLeuSerIleThrSerLeuGluAla	
	PheThrGlyGluSer	
	(cleavage site unknown)	

References are indicated in parentheses.
[a] D. Oxender (personal communication).

postulated that a threshold value of 18 Å for the hydrophobic axis length is critical for the export process. The threshold hydrophobic axis length can be achieved with 12 amino acids in α-helix structures or with five residues in β sheet structures (44, 45). The periodic structure of α-helix or β sheet followed by β turn near the cleavage site is consistent with the loop model postulated by Inouye and coworkers (17). Likewise, Steiner and coworkers have postulated a β transorption hypothesis that invokes β sheet structures in signal peptides to be important structural determinants in protein export (46). While β sheet structures have been predicted for the hydrophobic cores of certain signal sequences based on Chou-Fasman calculations, other theoretical considerations have led to the conclusion that α-helical conformations are more compatible with a nonpolar environment of the lipid bilayer (47).

Statistical analysis of signal sequences has also been applied to define a common sequence recognized by signal peptidase. Von Heijne, based on the analysis of 78 eukaryotic signal sequences, has concluded that signal sequences contain two different and largely independent domains (48). Accordingly, one domain forms a hydrophobic core spanning the membrane as a helix-plus-sheet structure, and it is presumably involved in the initiation of export through its interaction with SRP; the second domain resides in the region of −5 to −1 from the cleavage

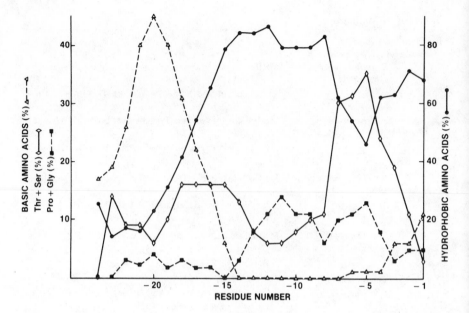

Fig. 3.3. Amino acid distribution in the signal peptides of prokaryotic precursor proteins. Using the signal sequences listed in Table 3.1, the content (C) of amino acid residues of a particular class at each position (N) is calculated according to Halegoua and Inouye (17), C=100×(total number of a particular class of amino acids at positions N−1, N and N+1)/(total number of amino acid residues at positions N−1, N, and N+1). C for each class is plotted against residue number from the cleavage site. (△), Contents of basic amino acid residues (Lys, Arg, and His); (◇), contents of Thr and Ser; (■), contents of Pro and Gly; and (●), contents of hydrophobic residues (Ala, Cys, Val, Met, Ile, Leu, Tyr, Phe, and Trp except for Met at the NH₂ termini).

site, and it confers the specificity of processing by signal peptidase. The latter structure appears to show strong preference for small neutral residues in both position −1 and position −3, which in turn define the cleavage site between −1 and +1 (Figure 3.4A).

A more detailed analysis was carried out by Perlman and Halvorson, based on 39 precursor proteins from diverse prokaryotic and eukaryotic sources (49). In addition to the finding of a hydrophobic core of approximately 12 amino acids preceded by positively charged residues in a typical signal sequence, Perlman and Halvorson noticed a nonrandom distribution of hydrophobic residues in the core. Following the hydrophobic cores that are aligned by the last residue, a highly nonrandom distribution of Ala was observed within the sequence following the core, with Ala strongly preferred at +4 and +6 positions, but not at +5, and the cleavage site at the +6 position from the end of the core. As can be seen in Figure 3.5, a nonrandom distribution of Ala in the signal sequences from prokaryotic precursor proteins listed in Table 3.1 is readily discerned. These observations strongly suggest the Ala-X-Ala sequence as the most frequent sequence preceding signal peptidase cleavage. Therefore, it has been proposed that a signal peptidase

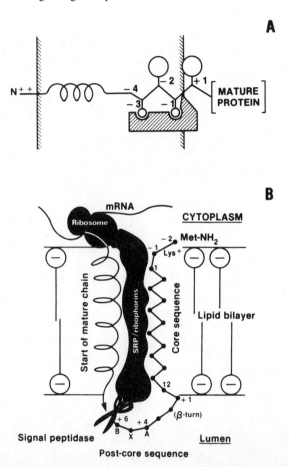

Fig. 3.4. Signal peptidase recognition site and sequence in eukaryotic and prokaryotic signal peptides. (A) Proposed signal sequence-signal peptidase complex according to von Heijne (48). The signal sequence spans the membrane as a "helix + sheet" structure and the small, neutral residues in position −1 and −3 fit into a pocket in the protease, thereby defining the cleavage site between positions −1 and +1. (Taken from [48].) (B) Proposed orientation of a typical nascent secretory polypeptide during protein secretion and processing. The transmembrane protein complexes including the SRP (signal recognition particle) and ribophorins are postulated to be involved in the signal sequence recognition, binding of nascent chains to the membrane, and insertion of nascent chains into the membrane followed by binding to the signal peptidase. The amino acid residues are aligned at position 1 by the first residue of the hydrophobic core. −1 and −2 refer to precore residues and +1 to +6 refer to postcore residues. The A-X-B sequence defines the sequence immediately preceding the signal peptidase cleavage site. (Taken from [49].)

recognition sequence of A-X-B (with Ala strongly preferred as A and B, but not as X) is located between the fourth and sixth amino acid following the core sequence (Figure 3.4B). Inasmuch as this preferred cleavage site is aligned with the carboxyl, but not amino, terminal ends of the hydrophobic core, it is further

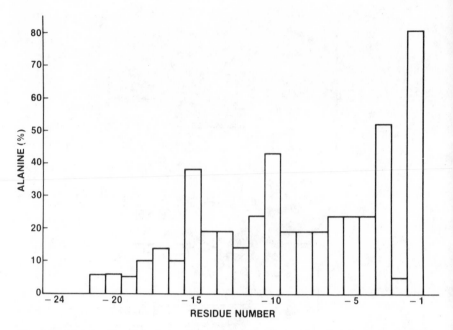

Fig. 3.5. Distribution of alanine residues in the signal peptides of prokaryotic precursor proteins listed in Table 3.1. The percent of alanine at each position is plotted against the residue number from the cleavage site.

postulated that the COOH termini of the core are oriented towards the signal peptidase located on the luminal surface of the ER. This orientation, together with the predicted β turn structure often found between the core and the cleavage site (Figure 3.4B), is consistent with the reverse (hairpin) insertion of the signal sequence proposed by several investigators (46, 50, 51).

A second approach to assess the structure-function relationship of signal peptides in protein export is the use of bacterial genetics. A large number of bacterial mutants altered in the signal sequences of exported proteins have been isolated and characterized (12, 52). These mutants fall into three general classes: 1) those selected for defective export; 2) those obtained without biased phenotype with regard to protein export; and 3) those constructed by *in vitro* mutagenesis in a systematic way to alter various segments of the signal peptides. Studies of these mutants have led to the following general conclusions.

1. The NH₂ terminal segment of signal peptide is important for the synthesis of the exported protein and for the initial association of the signal sequence with the putative export machinery. Positively charged amino acid residues are not absolutely required for the export of outer membrane lipoprotein in *E. coli*, but the replacement of these positively charged residues with neutral amino acids results in a reduced synthesis of this (53). Increased negative charge in this segment of prolipoprotein signal sequence also affects the rate of export,

presumably by altering the association of signal sequences with the export machinery in the membrane. Similar results were obtained with a particular *lam*B mutant allele affecting the NH_2 terminal hydrophilic portion of the λ receptor signal sequence (54).

2. The overproduction lethality phenotype of *E. coli* cells harboring hybrid proteins (consisting of NH_2 terminal *lam*B or *mal*E gene products fused with COOH terminal β-galactosidase) has presented a powerful selection for mutants defective in protein export. The distributions of intragenic mutations affecting protein exports obtained by this selection are nonrandom (12, 32, 55, 56). All of the mutations thus far isolated for defective export are mainly clustered in the region encoding the hydrophobic core of the signal peptide, and the mutational alterations are of three kinds: 1) substitution of neutral and hydrophobic residues with charged and hydrophilic residues; 2) deletion within the hydrophobic core and deletion extending from the core into the mature protein; and 3) amino acid substitution affecting secondary structures of the signal peptide. These results provide the strongest evidence for an essential role of the signal peptide in protein export. While the mutations affecting protein export are clustered within the hydrophobic segment of the signal peptide, the distribution of these mutations within the signal peptide is non-random, suggestive of the presence of specific residues in the signal peptide as the recognition site(s) by the putative SRP. Thus, each of the substitutions of Val-14 by Asp-14, Ala-15 by Glu-15, Ala-16 by Glu-16, and Met-19 by Arg-19 or Lys-19 in the signal peptide of LamB precursor protein results in a defective export. In contrast, the replacement of Gly-17 by Arg-17 or by Lys-17 does not affect export of LamB protein (57). These results taken together strongly suggest a general requirement for both an uninterrupted hydrophobic segment in the signal peptide that facilitates its spanning the cytoplasmic membrane and a subset of amino acid residues in the signal peptide that may participate in a specific interaction with the putative export machinery in *E. coli*. The postulated requirement of a threshold hydrophobic axis length of the signal peptide is supported by the isolation of second-site revertants of *lam*B signal sequence mutation that restore the function of the mutated signal peptide, presumably by a compensatory change in the secondary structure of the signal peptide (58, 59).

Further support for the interaction of the signal peptide with a putative SRP is provided by the isolation of extragenic suppressor mutations that restore the export of mutant proteins containing alterations in the signal peptide (12, 60–62). Furthermore, one of the suppressor genes, the *prl*A, appears to be essential for protein secretion in general, since a mutant with a specific allele (*sec*Y) of this gene is defective in the export of many outer membrane and periplasmic proteins when the mutant cells are grown at the nonpermissive temperature (63, 64). The dual phenotype of the *sec*Y/*prl*A mutants (*i.e.*, suppression of signal sequence mutation and a defect in protein export in general) are strongly suggestive of a specific interaction between the SecY protein and signal peptides of exported proteins.

Functions of Signal Peptides in Protein Export

There are at least two distinct functions that are encoded in the primary structures of signal peptides; *i.e.*, the initiation of export process through an interaction of the signal sequence with the export machinery, and the completion of the export process through removal of the signal peptides by signal peptidases. The role of the signal peptide, if any, in the transmembrane translocation of the precursor protein, following the association of SRP with the membrane receptor (docking) protein, remains totally obscure.

That the signal sequence is necessary for the initiation of protein export is based mostly on genetic studies of protein export in *E. coli*. The extreme preponderance of mutations with alterations in signal peptides, in contrast with the near absence of mutations affecting the mature protein, among mutants selected for defective protein export provides the strongest evidence for the role of the signal sequence in the cotranslational model of protein secretion as envisioned by the signal hypothesis (65–67). It is equally clear, however, that the signal sequence is not sufficient for directing a protein to be secreted through the membrane (68–70). Hybrid proteins containing intact signal peptides of exported proteins, plus a limited number of additional amino acid residues of the mature proteins linked to otherwise cytosolic proteins, remain localized in the cytoplasm. These results suggest that additional structural information that resides in the mature protein is required for the extracellular localization of the exported proteins. Whether the cytoplasmic localization of these hybrid proteins results from a failure to initiate export, or is a secondary consequence of the release of the hybrid protein from the membrane following abortive export, is not known.

A second function of the signal sequence is to provide the cleavage site for the processing of precursor proteins by signal peptidases. Once again, the use of bacterial mutants, altered in the structures of the exported proteins, has provided new insights as to whether the signal peptide contains both necessary and sufficient information for the correct processing of precursor proteins by signal peptidases (12, 70, 71–75).

A related question concerns the number and substrate specificities of signal peptidases. Existing data would appear to favor the existence of a limited number of signal peptidases with relatively broad substrate specificity, with regard to the total structure of the precursor or mature proteins. Thus, *E. coli* signal peptidase is capable of processing precursor proteins encoded by genes from eukaroytic cells, and the converse is true for the signal peptidase in the ER of dog pancreas, which correctly processes precursor proteins from bacteria or bacteriophage (76–78). The diverse primary structures of signal sequences in prokaryotic and eukaryotic cells suggest the possibility that signal peptidases recognize a localized structure near the cleavage site in the signal peptide. This notion is supported by the statistical analysis of known signal sequences and by the characterization of bacterial mutants altered in the structures of the exported proteins that are defective in the processing of the precursor proteins by the signal peptidase. It is clear that signal peptidase, which is an endopeptidase, does not measure the length of the signal peptide from its NH_2 terminus as a mechanism for determining the appropriate cleavage site (76, 79). The preference of small uncharged residues near the cleavage site, with a strong bias towards the sequence Ala-X-Ala immediately preceding the cleavage site (48, 49), is consistent with the notion that the signal peptidase recognizes a localized consensus sequence near the cleavage site. The other two features common to many different signal peptides (*i.e.*, a secondary structure of β turn near the cleavage site and the nearly constant distance [six amino acid residues] between the cleavage site and the COOH end of the hydrophobic segment) may constitute part of the common structures recognized by the signal peptidase (49).

The recognition site for the processing enzyme appears to be distinct from that for the export machinery, since mutations affecting export do not necessarily alter processing by signal peptidase when the export defect is suppressed by unlinked suppressor mutations (60). Likewise, mutants altered in the *bla* gene have been isolated that contain signal sequence mutations affecting the processing, but not the export, of pro-βlactamase (72). Results obtained from the studies of signal sequence mutations of *lam*B and *bla* clearly indiate that the secondary structures of four amino acid residues within the signal sequence adjacent to the site of cleavage, constitute an important recognition site for the signal peptidase.

While the signal peptide most likely contains the crucial recognition site for the signal peptidase, structural alterations in the mature protein also appear to affect the cleavage reaction. An in-frame deletion in the *lam*B gene removing amino acids 70 through 200 of the mature protein prevents processing of the precursor form of the mutant LamB protein by the signal peptidase (70). A mutation that results in the substitution of the glutamic acid by leucine at the second amino acid residue of the mature M13 major coat protein also affects the processing of the mutant precursor protein both *in vivo* and *in vitro* by signal peptidase (71). A mutation in the *lpp* gene located near the COOH terminus of prolipoprotein also appears to affect the processing of the mutant prolipoprotein by signal peptidase (80). Thus, the recognition site for the processing enzyme appears to extend beyond the signal sequence.

That both the local structures of the signal peptides and the overall conformation of the precursor proteins contribute to the specificity of processing reaction is also made evident by studies of mutant precursor proteins and hybrid proteins. It has been recently shown that there are at least two signal peptidases in *E. coli* with distinct substrate specificities (see below) (81, 82). Signal peptidase I (SPase I) is presumably responsible for processing precursor proteins with Ala-X-Ala↓ preceding the cleavage site, while signal peptidase II (SPase II) recognizes Leu-Ala-Gly↓-(glyceride-cysteine) in modified prolipoproteins as the cleavage site (Table 3.2) (18, 40, 41, 83–86). In mutants synthesizing unmodified prolipoproteins due to mutations in the *lpp* gene, SPase I does not cleave these prolipoproteins, even though they contain potential cleavage sites for SPase I (87, 88). In a particular mutant constructed by site-directed mutagenesis, replacement of Cys with Gly results in the accumulation of unmodified and unprocessed prolipoprotein in the mutant cells. Apparently, this unmodified precursor protein is not

Table 3.2. Signal Sequences of Lipoprotein Precursors in Bacteria

Murein lipoprotein
  E. coli (18)   −20
                            MetLysAlaThrLysLeuValLeuGlyAlaValIleLeuGlySerThrLeu
                                        +1
                            *LeuAlaGlyCys*SerSerAsnAla

  S. marcescens (83)[a]   −20
                            MetAsnArgThrLysLeuValLeuGlyAlaValIleLeuGlySerThrLeu
                                        +1
                            *LeuAlaGlyCys*SerSerAsnAla

  E. amylovora (84)   −20
                            MetAsnArgThrLysLeuValLeuGlyAlaValIleLeuGlySerThrLeu
                                        +1
                            *LeuAlaGlyCys*SerSerAsnAla

  M. morganii (85)   −20
                            MetGlyArgSerLysIleValLeuGlyAlaValValLeuAlaSerAlaLeu
                                        +1
                            *LeuAlaGlyCys*SerSerAsnAla

  P. mirabilis[a]   −19
                            MetLysAla---LysIleValLeuGlyAlaValIleLeuAlaSerGlyLeu
                                        +1
                            *LeuAlaGlyCys*SerSerSerAsn

Penicillinase
  B. licheniformis (40)   −26
                            MetLysLeuTrpPheSerThrLeuLysLeuLysLysAlaAlaAlaValLeuLeuPhe
                                        +1
                            SerCysValAla*LeuAlaGlyCys*AlaAsnAsnGln

  S. aureus (41)   −16   +1
                            MetLysLysLeuIlePheLeuIleValIleAlaLeuVal*LeuSerAla*CysAsnSerAsnSer

TraT protein (86)   −20
                            MetLysMetLysLysLeuMetMetValAlaLeuMetSerSerThrLeuAla
                                        +1
                            *LeuSerGlyCys*GlyAlaMetSer

  Consensus sequence   +1
                            *LeuAlaGlyCys*SerSerAsn
                              SerAla GlyAlaMet
                                 AlaAsnSer
                                 Asn

References are indicated in parentheses.
[a] M. Inouye (personal communication).

processed by SPase I to any appreciable extent. When the signal peptide and first nine amino acid residues of this mutant prolipoprotein (Cys-21 → Gly-21) are fused with the mature protein of β-lactamase, the hybrid protein appears to be processed (M. Inouye, personal communication). The same construction, with wild-type prolipoprotein signal peptide plus nine amino acids and β-lactamase, results in the formation of modified and processed lipo-β-lactamase, as expected (89). This result suggests that the overall conformations of the precursor proteins may play an important role in the formation of the recognition site for signal peptidase. A similar conclusion was reached in a study using *Bacillus licheniformis* penicillinase mutants, which contain specific alterations in the modification and cleavage site in the signal peptide (75). A mutant with a pentapeptide (Leu-24Ala-25Gly-26Cys-27Ala-28) deletion spanning the cleavage site in the signal peptide of *B. licheniformis* penicillinase is shown to synthesize unmodified and unprocessed prepenicillinase as the major species of the *pen* gene product. On the other hand, a point mutation in the *pen* gene with the replacement of Cys-27 by Ser-27 results in the formation of processed penicillinase; the processing occurs either at Ala-28 (Gly-Ser-Ala↓) or at Ala-34 (Thr-Asn-Ala↓). Apparently, these potential-cleavage sites are not exposed in the deletion mutant to allow processings to occur at these sites, presumably by distinct signal peptidase(s).

The notion that tertiary structures of precursor proteins may contribute to the recognition of signal peptide by signal peptidase appears to correlate well with the timing of the processing reaction, relative to the extent of completion of the synthesis of the precursor proteins. The work of Randall and coworkers has clearly demonstrated that the processing of cotranslationally secreted proteins is a late event, and it does not take place until the synthesis of the precursor proteins has reached 80% of the full-size polypeptide (90). Randall has further shown that the transmembrane export of nascent chains of secretory proteins does not take place until the nascent chain has reached a critical size, which is also approximately 80% of the full-size precursor protein (15). These results suggest that by the time the nascent chains become accessible to the signal peptidase, the synthesis of these precursor proteins have reached 80% of completion. It is therefore quite reasonable that mutations far removed from the cleavage site may affect the processing of precursor proteins by signal peptidases, presumably by altering the conformations of the precursor proteins.

Consistent with the notion that the processing of precursor proteins by signal peptidase is a late event in the overall process of protein secretion, it has been shown that processing is not required for the transmembrane export of the secretory proteins. Mutant precursor proteins that are not cleavable by signal peptidases are exported (60, 62, 87, 88). Similar conclusions have been reached for the secretion of ovalbumin (which does not contain cleavable signal peptide) (91) and the secretion of preprolactin containing amino acid analogs (which interfere with the processing) (92).

What, then, is the function of signal peptidase if processing is not required for protein secretion? The available evidence suggests the cleavage of precursor protein by signal peptidase as a final step of the secretion process, thereby releasing the already translocated polypeptides from the membrane into the lumen of the ER of the eukaryotic cell, into the culture medium of gram-positive bacteria (93–96), or into the periplasm of gram-negative bacteria (72). As to outer membrane proteins of gram-negative bacteria, uncleaved precursor forms of outer membrane proteins may be assembled into the outer membrane via the mature polypeptides, while the uncleaved signal peptides remain anchored in the cytoplasmic membrane. In this view, the assembly of proteins into the outer membrane is a separate step distinct from translocation and processing; and the removal of signal peptide is essential for the complete separation of outer membrane proteins from the cytoplasmic membrane. Studies on the distribution of uncleaved prolipoprotein in *E. coli* provide further evidence for the model stated above. Uncleaved but glyceride-modified prolipoprotein is translocated across the cytoplasmic membrane, since the prolipoprotein can be shown to be covalently linked to the peptidoglycan (97). On the other hand, the apparent subcellular localization of the prolipoprotein appears to vary with the experimental procedures used (sucrose density gradient centrifugation vs. Sarkosyl solubilization) and the nature of the prolipoprotein (modified vs. unmodified). The location of glyceride-modified prolipoprotein was assigned to the cytoplasmic membrane fraction based on sucrose density gradient centrifugation (98). Glyceride-modified prolipoprotein was found to be insoluble in Sarkosyl, which is suggestive of a tight association with the outer membrane (99). In contrast, unmodified prolipoproteins, containing various point mutations in the signal peptide affecting the modification reaction, appeared to be associated with the outer membrane, based on both sucrose density gradient centrifugation and Sarkosyl solubilization, but the degree of association of the mutant prolipoproteins with the outer and inner membranes varied with different mutant proteins (73, 87, 88, 100). Even mutants with unmodified prolipoprotein mainly localized in the outer membrane appeared to suffer from the presence of these unprocessed precursor proteins in the cell (88); such overproduction lethality is usually seen when the cytoplasmic membrane is jammed with hybrid proteins (101). One possible explanation for these observations is that the uncleaved precursor protein interacts both with the cytoplasmic membrane via the signal peptide and with the outer membrane via the mature protein, which has been translocated and assembled into the outer membrane. The hydrophobicity of the prolipoprotein molecules (lipid modified, introduction of charged residues within the signal peptide due to mutation, etc.) may affect the partition of prolipoprotein into the cytoplasmic and outer membranes, and may also affect the secondary association of these molecules with these membranes during subcellular fractionation.

This interpretation is consistent with the notion that proteolytic cleavage of the precursor protein is not a rate-limiting step in protein secretion. Nevertheless, it could be a kinetically important step for the release of nascent chains from the membrane into the aqueous environment. The strategic location of this sequence at the NH_2 terminus of an exported protein serves a dual purpose of facilitating a cotranslational initiation of the export process and allowing the release of the mature protein from the membrane. For different exported proteins, the primary structure of the signal peptides are hardly conserved. Yet, the conservation of such a signal sequence as a transient structure at the NH_2 terminus during the export of a structurally diverse group of proteins, strongly favors an important role for proteolytic cleavage in protein export.

Signal Peptidases

The existence of signal peptidase activity in the ER membrane as part of the secretory machinery was first postulated as a tenet of the signal hypothesis, and it was subsequently demonstrated in dog pancreas microsomal membranes (102, 103). Similar activities have been found in rat liver microsomes (104), mitochondrial matrix from yeast (105–107), rough ER from hen oviduct (108, 109), and microsomal membranes from *Drosophila melanogaster* embryos (110). None of these enzymes has been purified and characterized. Preliminary results would suggest that these enzymes behaved like integral membrane proteins and were not readily extractable with low concentrations of detergents, EDTA, or hight salts, nor with carbonate buffer at pH 11.5 (108). An apparent requirement of phospholipids has been reported for the activity of signal peptidase localized in dog pancreas microsomes (111). Phosphatidylcholine was found to be most effective and phosphatidylinositol less effective, whereas phosphatidylserine, phosphatidylethanolamine, sphingomyelin, and lysophosphatidylcholine were all ineffective in restoring activity of delipidated canine pancreatic signal peptidase (111).

Lively and Walsh have undertaken a detailed study of hen oviduct signal peptidase (108, 109). The enzyme was solubilized from hen oviduct rough ER membrane with octyl-β-D-glucopyranoside. The cleavage of the precursor of human placental lactogen (pre-hPL) was found to be dependent upon both a unique conformation of the substrate (which could be induced by the inclusion of an optimal concentration of anti-hPL during the synthesis of pre-hPL) and the relative concentrations of detergent and membrane phospholipids. A molar ratio of detergent-to-phospholipid greater than 60 resulted in an apparent inhibition of the signal peptidase activity. Therefore, the apparent requirement of signal peptidase activity for phospholipids may be explained in terms of a change in the physical state of the substrate, rather than that of the enzyme, at high detergent concentrations. Inasmuch as the assay system involves mixed micelles, the apparent inhibition of signal peptidase activity by high detergent concentrations, and its reversal by phospholipids, may simply be a reflection of the partition of substrate into the detergent micelles that do not contain the signal peptidase. Attempts to extract this enzyme from the membrane vesicles prepared from rough ER of hen oviduct revealed a very tight association of this enzyme with the membrane vesicles. Even after extraction with low concentrations of detergent (which do not dissolve the lipid bilayer) or with alkaline carbonate buffer (which releases most of the membrane-associated proteins), signal peptidase remained active in the residual membrane vesicles. These results indicate that this enzyme possesses the characteristics of an integral membrane protein.

The difficulty in obtaining pure signal peptidase may also be attributed to the low amounts of this enzyme, as evidenced by the recent reports on the purification and characterization of two signal peptidases from *E. coli* (Table 3.3). It is perhaps not too surprising that purification to apparent homogeneity and characterization of M13 procoat protein signal peptidase (SPase I) was achieved by Wickner and coworkers following the cloning of its structural gene (*lep*) and the amplification of its expression (112–114).

The purified SPase I has a subunit molecular weight of 37,000, which is in good agreement with the predicted molecular weight of 35,994 based on the amino acid sequence deduced from the nucleotide sequence of the *lep* gene (115). This enzyme appears to be active in its monomeric form, and it is present in *E. coli* cells in low amounts (approximately 500 molecules per cell). The subcellular localization of this enzyme in both inner and outer membrane fractions of the *E. coli* cell envelope was also demonstrated (116). This observation raises the interesting possibility that the dual localization of this enzyme may be responsible for the processing of precursors of most, if not all, periplasmic and outer membrane proteins during their export. On the other hand, recent results based on studies using *E. coli* strains overproducing SPase I seem to indicate that most of the overproduced SPase I appears to reside in the cytoplasmic membrane fraction (115). SPase I appears to be a peripheral membrane protein; the hydrophobicity of this protein (41.5% polarity) is similar to those of water-soluble globular

Table 3.3. Comparison of Signal Peptidase I and Signal Peptidase II of *Escherichia coli*

Properties	SPase I		SPase II
Structural gene and map position	*lep*, 55 min		*lsp*, 0.5 min
Subunit molecular weight	36,000		18,000
Subcellular localization	Inner and outer membrane		Inner membrane
Topography	Peripheral membrane protein at the external surface of inner membrane		Transmembrane integral membrane protein
Inhibitors			
Globomycin	Insensitive		Sensitive
HgCl$_2$, NEM	N.D.[a]		Sensitive
Substrate and cleavage sites	M13 procoat protein	Ala-Ala	Prolipoprotein Gly-GlycerideCys
	LamB precursor	Ala-Val	Prepenicillinase Gly-GlycerideCys
	MBP precursor	Ala-Lys	PreTraT protein Gly-GlycerideCys
	OmpA precursor	Ala-Ala	
	LSBP precursor	Ala-Asp	
Essential function	Yes		Yes

[a] N.D., Not determined.

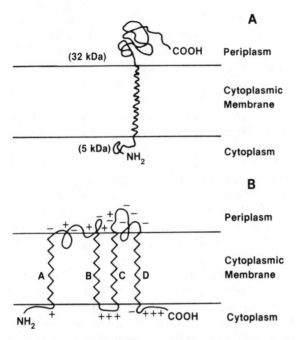

Fig. 3.6. Hypothetical transmembrane structures of the M13 procoat protein signal peptidase (SPase I) and prolipoprotein signal peptidase (SPase II). (A) Anchorage of SPase I via its NH_2 terminal segment. (Based on the data in [115].) (B) Transmembrane structure of SPase II (as postulated in [121].) The four hydrophobic segments (A, residues 12 to 22; B, 74 to 86; C, 100 to 109; and D, 142 to 153) are shown here as β sheet structures embedded in the cytoplasmic membrane. The relative locations of charged residues (Arg, Lys, Asp, and Glu) are shown with "+" or "−" signs. (For the sequence and hydropathy profile of SPase II, see [121].)

proteins. A large portion of this enzyme appears to be localized to the outer surface of the cytoplasmic membrane, and it is accessible to proteolytic digestion. When intact spheroplasts were treated with trypsin, 10% of the SPase I polypeptide was protected from protease by its integration into the membrane; data from tryptic peptide mapping of the intact SPase I and tryptic-resistant fragment suggest that SPase I was anchored in the inner membrane by its NH_2 terminus (Figure 3.6A) (115).

It has been reported that in addition to M13 procoat protein, purified SPase I can process the precursor proteins of a number of periplasmic and outer membrane proteins, including maltose-binding protein, λ receptor protein, and leucine-specific-binding protein. However, published data are available only for the processing of pre-OmpA (117), pre-MBP (118), and M13 procoat protein (119) by the purified SPase I. *In vivo* evidence establishes unambiguously that SPase I is responsible for the processing of M13 procoat protein, since the rate of maturation of a mutant M13 procoat protein was enhanced by the overproduction of SPase I activity (113). Similar results have not been available for other indigenous *E. coli*-exported proteins, and a definitive assessment of the role of SPase I in the processing of most *E. coli* precursor proteins other than procoat protein awaits the isolation of a conditionally lethal mutant defective in SPase I activity. Negative genetic evidence would tend to support that *lep* is, indeed, an essential gene in *E. coli* (120).

Biochemical and genetic evidence have led to the conclusion that the prolipo-protein signal peptidase (SPase II) is distinct from SPase I (81, 82). No sequence homology in either the nucleotide sequences of these two genes or the amino acid sequences of these two enzymes can be discerned (115, 121). The two unique properties characteristic of SPase II may be functionally related, *i.e.*, its exquisite sensitivity to inhibition by globomycin (a hydrophobic cyclic peptide) (122) and its requirement for glyceride-modified precursor protein as the substrate (81, 82, 123). SPase II activity is located in the inner membrane of the *E. coli* cell enve-lope (124, 125), and it has been purified extensively (123). While SPase I was purified about 6000 fold in order to reach apparent homogeneity, an approximate 35,000-fold purification of SPase II was needed to yield a highly purified enzyme with an apparent subunit molecular weight of $17,800 \pm 900$ (123). This is in excel-lent agreement with the molecular weight of 18,143 based on the amino acid sequence of SPase II deduced from the nucleotide sequence of the *lsp* gene (121). The molecular weight of the native enzyme remains to be determined. The puri-fied enzyme from *E. coli* B is inhibited by globomycin with an I_{50} of 36 nM (123), whereas the same enzyme in the crude membrane extract of *E. coli* K12 is inhibited by globomycin, with I_{50} varying from 0.7 nM to 7 nM depending on the nature of the substrate (124, 126). These differences might be attributed to differ-ences in experimental conditions, especially the source of the enzyme and the concentration and nature of the substrate. SPase II may be a thiol protease, since it is inhibited by $HgCl_2$ and N-ethylmaleimide and requires dithiothreitol for sta-bility (123). It is not inhibited by the known serine protease inhibitors or the known metalloendo- or exoprotease inhibitors. It is an endopeptidase, exhibiting a unique substrate specificity. While glyceride-modified prolipoprotein is effi-ciently cleaved at the correct site between glycine and modified cysteine by the purified enzyme to yield the processed lipoprotein and signal peptide in stoichio-metric quantities, neither the unmodified prolipoprotein nor the precursor of maltose-binding protein is processed by purified SPase II (123). The specific requirement for glyceride-modified cysteine at the cleavage site of the substrate and the apparent specific inhibition by the hydrophobic peptide antibiotic globo-mycin are consistent with the notion that this enzyme contains hydrophobic domains interacting with the lipid moiety (121).

The unique sensitivity of SPase II towards globomycin was exploited in the cloning of the gene encoding SPase II (127). The same gene (*lsp*) was cloned independently through genetic complementation of an *E. coli* mutant that was deficient in SPase II activity, which was obtained in multiple steps including a selection for resistance to globomycin (128, 129). The *lsp* gene has been mapped (130, 131) and the nucleotide sequence determined (121). The DNA sequence of

```
        1                      10                                    20
1.  Met Ala Asn Met Phe Ala Leu Ile  Leu Val Ile Ala Thr Leu Val Thr Gly Ile Leu Trp
                              30
    Cys Val Asp Lys Phe Phe Phe Ala Pro Lys

        1                      10                                    20
2.  Met Ser Gln Ser Ile Cys Ser Thr Gly Leu Arg Trp Leu Trp Leu Val Val Val Val Leu
                              30
    Ile Ile Asp Leu Gly Ser Lys Tyr Leu Ile
```

Fig. 3.7. Amino terminal sequences of signal peptidase I (115) and signal peptidase II (121).

the *lsp* gene predicts a single polypeptide of 18 kd, comprising 164 amino acid residues. Like SPase I, SPase II does not contain an apparent signal peptide at its NH$_2$ terminus (Figure 3.7) (115, 121). Computer-assisted secondary structure analysis of the deduced amino acid sequence identified four hydrophobic regions that share features common to transmembrane segments in other integral membrane proteins. The hydrophobicity of this protein (29.9% polarity) is much higher than that of SPase I (41.5% polarity) (Table 3.4) (115, 121). Each of the four hydrophobic domains contains at least 10 amino acid residues and corresponds to regions predicted to be in β sheet conformation sufficient to span the lipid bilayer of the *E. coli* inner membrane. One possible transmembrane structure of SPase II is shown in Figure 3.6B. This model provides an explanation for the reduced SPase II activity in mutants containing deletions that remove positively charged COOH terminal residues (121).

Like the *lep* gene, which is located at 55 min on the *E. coli* genetic map (132), the *lsp* gene at 0.5 min (130, 131) is an essential gene. This is evidenced by the bactericidal activity of globomycin (122) and by the isolation of the conditionally lethal mutant possessing a defective *lsp* gene (129). It remains to be seen whether the *lsp* gene is essential in the absence of *lpp* gene.

Studies on the genomic organization of the *lsp* gene have revealed unexpected findings. The *lsp* gene is located downstream from *ile*S (encoding the isoleucyl-tRNA synthetase), and the terminating codons (UG*AUGA*) in the *ile*S gene overlap with the initiating codon of the *lsp* gene (Figure 3.8) (121). The overlapping of the stop-start codons at the cistronic-junctions of *ile*S and *lsp* suggests the interesting possibility that the expression of these two genes, having no apparent physiological relationship, may be translationally coupled. The *lep* gene is also separated from its promoter by a gene whose product and function remain to be established (113). The organization of *lep* and *lsp* genes as part of transcriptional units containing other genes (113, 120, 121) raises interesting questions regarding the regulation of their expression.

Both SPase I and SPase II appear to lack NH$_2$ terminal signal peptides and are assembled into the cytoplasmic membrane. The mode of assembly of SPase I remains an interesting enigma, since a large portion of this enzyme appears to be located at the outer (periplasmic) surface of the cytoplasmic membrane (115).

Table 3.4. Amino Acid Composition and Mole Percent Polarity of Signal Peptidase I and Signal Peptidase II from *Escherichia coli*

Amino Acid	SPase I	SPase II
Lys	19	4
His	3	3
Arg	17	6
Cys	3	2
Asp	17	8
Asn	10	7
Thr	18	5
Ser	17	10
Glu	18	1
Gln	15	5
Pro	23	3
Gly	27	15
Ala	25	18
Val	23	12
Met	8	4
Ile	22	14
Leu	24	23
Tyr	11	6
Phe	17	12
Trp	6	6
Total	323	164
Polarity (mole%)	41.5	29.9

The high overall hydrophobicity of SPase II, as well as regions of localized hydrophobicity in the enzyme, suggests that this enzyme may be assembled into the cytoplasmic membrane like many integral membrane proteins, such as the lactose carrier protein.

Concluding Remarks

Significant progress has been made in the study of the structure and function of signal peptides in bacteria. Mutant analysis has established an essential role of the signal peptide in protein export. Structural features in the signal peptides that are important for optimal synthesis of the exported proteins, for the interactions of signal peptides with the export machinery, and for the recognition by signal peptidases have been delineated. Much less is known regarding the mechanism by which the nascent chains are translocated, either cotranslationally or posttranslationally, across the membrane (16). The studies on the mechanism of transmembrane transfer of a large domain of nascent precursor proteins are related to the question of the energetics of the transfer process (133–135).

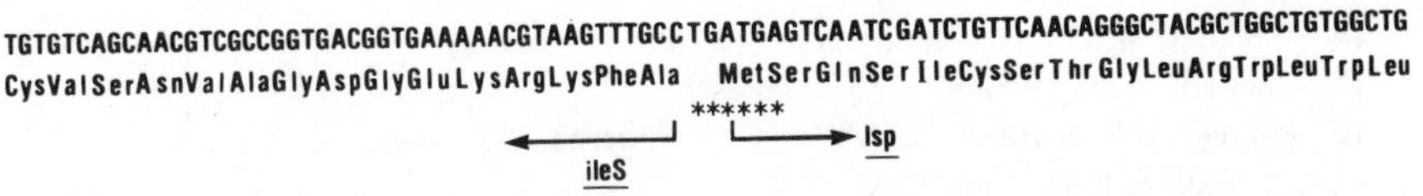

TGTGTCAGCAACGTCGCCGGTGACGGTGAAAAACGTAAGTTTGCCTGATGAGTCAATCGATCTGTTCAACAGGGCTACGCTGGCTGTGGCTG
CysValSerAsnValAlaGlyAspGlyGluLysArgLysPheAla   MetSerGlnSerIleCysSerThrGlyLeuArgTrpLeuTrpLeu

Fig. 3.8. Overlapping stop-start codons at the cistronic junction of *ileS* and *lsp*. The 15 amino acid COOH terminal sequences of isoleucyl-tRNA synthetase (140) and the NH₂ terminal sequence of SPase II (121) are shown.

Rapid progress has been made with regard to the purification and characterization of signal peptidases from *E. coli*; and progress in the characterization of signal peptidases from other prokaryotic or eukaryotic cells can be anticipated. Work similar to the pioneering studies of Blobel, Walter, Dobberstein, and Meyer, in their identification of SRP and docking protein in eukaryotic cells, is being actively pursued in prokaryotic cells. It is hopeful that the genetic studies of Beckwith, Silhavy, Ito, and Bassford will be corroborated by the biochemical approaches pioneered by Davis and Tai and their coworkers (136–138).

Besides the biochemical and genetic identification and characterization of the components involved in the protein export process, studies of the regulation of genes required for protein secretion constitute yet another new avenue for future investigations. The autoregulation of the synthesis of the SecA protein (139) and the genetic organization of *lep* and *lsp* genes as part of operons containing functionally unrelated genes (113, 120, 121), represent two examples of unexpected findings.

Acknowledgments.   The work from the author's laboratory has been supported by grants from United States Public Health Service grant GM-28811 and American Heart Association grant 81-663. Special thanks are due Dr. Paul Rick for critical reading of the manuscript.

References

1. Blobel, G. & Dobberstein, B. (1975) J. Cell Biol. *67*, 835–851.
2. Walter, P., Ibrahimi, I. & Blobel, G. (1981) J. Cell Biol. *91*, 545–550.
3. Walter, P. & Blobel, G. (1981) J. Cell Biol. *91*, 551–556.
4. Walter, P. & Blobel, G. (1981) J. Cell Biol. *91*, 557–561.
5. Walter, P. & Blobel, G. (1982) Nature *299*, 691–698.
6. Gilmore, R., Blobel, G. & Walter, P. (1982) J. Cell Biol. *95*, 463–469.
7. Gilmore, G., Walter, P. & Blobel, G. (1982) J. Cell Biol. *95*, 470–477.
8. Gilmore, R. & Blobel, G. (1983) Cell *35*, 677–685.
9. Meyer, D.I. & Dobberstein, B. (1980) J. Cell Biol. *87*, 503–508.
10. Meyer, D.I., Krause, E. & Dobberstein, B. (1982) Nature *297*, 647–650.
11. Meyer, D.I. (1982) TIBS *Sept.*, 320–321.
12. Silhavy, T.J., Benson, S.A. & Emr, S.D. (1983) Microbiol. Rev. *47*, 313–344.
13. Wickner, W. (1979) Ann. Rev. Biochem. *48*, 23–45.

14. Wickner, W. (1980) Science *210*, 861-868.
15. Randall, L.L. (1983) Cell *33*, 231-240.
16. Randall, L.L. & Hardy, S.J.S. (1984) In Modern Cell Biology, vol. 3, 1-20, Satir, B., ed. Alan R. Liss, New York.
17. Halegoua, S. & Inouye, S. (1979) In Bacterial Outer Membrane. Biogenesis and Function, 67-113, Inouye, M., ed. John Wiley & Sons, New York.
18. Nakamura, K. & Inouye, M. (1979) Cell *18*, 1109-1117.
19. Movva, N.R., Nakamura, K. & Inouye, M. (1980) J. Biol. Chem. *255*, 27-29.
20. Braun, G. & Cole, S.T. (1982) Nucl. Acids Res. *10*, 2367-2378.
21. Hedgpeth, J., Clement, J.M., Marchal, C., Perrin, D. & Hofnung, M. (1980) Proc. Natl. Acad. Sci. USA *77*, 2621-2625.
22. Overbeeke, N., Bergmans, H., Mansfeld, F.V. & Lugtenberg, B. (1983) J. Mol. Biol. *163*, 513-532.
23. Mutoh, N., Inokuchi, K. & Mizushima, S. (1982) FEBS Lett. *137*, 171-174.
24. Mizuno, T.M., Chou, Y. & Inouye, M. (1983) FEBS Lett. *151*, 159-164.
25. Inouye, H., Barnes, W. & Beckwith, J. (1982) J. Bacteriol. *149*, 434-439.
26. Jaurin, B. & Grundstrom, T. (1981) Proc. Natl. Acad. Sci. USA *78*, 4897-4901.
27. Sutcliffe, J.G. (1978) Proc. Natl. Acad. Sci. USA *75*, 3737-3741.
28. Scripture, J.B. & Hogg, R.W. (1983) J. Biol. Chem. *258*, 10853-10856.
29. Higgins, C.F. & Ames, G.F.L. (1981) Proc. Natl. Acad. Sci. USA *78*, 6038-6042.
30. Oxender, D.L., Anderson, J.J., Daniels, C.J., Landick, R., Gunsalus, R.P., Zurawski, G. & Yanofsky, C. (1980) Proc. Natl. Acad. Sci. USA *77*, 2005-2009.
31. Landick, R. & Oxender, D.L. (1982) In Membranes and Transport, vol. 2, 81-88, Martonosi, A., ed. Plenum Press, New York.
32. Bedouelle, H., Bassford, P.J., Jr., Fowler, A.V., Zabin, I., Beckwith, J. & Hofnung, M. (1980) Nature *285*, 78-81.
33. Groarke, J.M., Mahoney, W.C., Hope, J.N., Furlong, C.E., Robb, F.T., Zalkin, H. & Hermodson, M.A. (1983) J. Biol. Chem. *258*, 12952-12956.
34. Sugimoto, K., Sugisaki, H., Okamoto, T. & Takanami, M. (1977) J. Mol. Biol. *110*, 487-507.
35. Shaller, H., Beck, E. & Takanami, M. (1979) In The Single Stranded DNA Phages, 139-153, Denhart, D., Dressler, D. & Ray, D., eds. Cold Spring Harbor Laboratory, Cold Spring Harbor, New York.
36. Dallas, W.S. & Falkow, S. (1980) Nature *288*, 499-500.
37. Spicer, E.K. & Noble, J.A. (1982) J. Biol. Chem. *257*, 5716-5721.
38. Palva, I., Pettersson, R.F., Kalkkinen, N., Lehtovaara, P., Sarvas, M., Söderlund, H., Takkinen, K. & Kääriänen, L. (1981) Gene *15*, 43-51.
39. Stephens, M.A., Ortlepp, S.A., Ollington, J.F. & McConnell, D.J. (1984) J. Bacteriol. *158*, 369-372.
40. Neugebauer, K., Sprengel, R. & Schaller, H. (1981) Nucl. Acids Res. *9*, 2577-2588.
41. McLaughlin, J.R., Murray, C.L. & Rabinowitz, J.C. (1981) J. Biol. Chem. *256*, 11283-11291.
42. Sloma, A. & Gross, M. (1983) Nucl. Acids Res. *11*, 4997-5004.
43. Chou, P.Y. & Fasman, G.D. (1978) Ann. Rev. Biochem. *47*, 251-276.
44. Bedouelle, H. & Hofnung, M. (1981) In Intermolecular Forces, 361-372, Pullman, B., ed. D. Reidel Publishing Co.
45. Bedouelle, H. & Hofnung, M. (1981) In Membrane Transport and Neuroreceptors, 339-403, Oxender, D., et al., eds. Alan R. Liss, New York.

46. Steiner, D.F., Quinn, P.S., Chan, S.J., Marsh, J. & Tager, H.S. (1980) Annals NY Acad. Sci. *343*, 1-16.
47. Pincus, M.R. & Klausner, R.D. (1982) Proc. Natl. Acad. Sci. USA *79*, 3413-3417.
48. Von Heijne, G. (1983) Eur. J. Biochem. *133*, 17-21.
49. Perlman, D. & Halvorson, H.O. (1983) J. Mol. Biol. *167*, 391-409.
50. Inouye, M. & Halegoua, S. (1980) CRC Crit. Rev. Biochem. *10*, 339-371.
51. Engelman, D.M. & Steitz, T.A. (1981) Cell *23*, 411-422.
52. Vlasuk, G.P., Ghrayeb, J. & Inouye, M. (1984) In The Enzymes of Biological Membranes, vol. 2, 309-329, Martonosi, A., ed. Plenum Press, New York.
53. Vlasuk, G.P., Inouye, S., Ito, H., Itakura, K. & Inouye, M. (1983) J. Biol. Chem. *258*, 7141-7148.
54. Hall, M., Gabay, N.J. & Schwartz, M. (1983) EMBO J. *2*, 15-19.
55. Emr, S.D., Hedgpeth, J., Clement, J.M., Silhavy, T.J. & Hofnung, M. (1980) Nature *285*, 82-85.
56. Emr, S.D. & Silhavy, T.J. (1980) J. Mol. Biol. *141*, 63-90.
57. Emr, S.D. & Silhavy, T.J. (1982) J. Cell Biol. *95*, 689-696.
58. Emr, S.D. & Silhavy, T.J. (1983) Proc. Natl. Acad. Sci. USA *80*, 4599-4603.
59. Bankaitis, V.A., Rasmussen, B.A. & Bassford, P.J., Jr. (1984) Cell *37*, 243-252.
60. Emr, S.D. & Bassford, P.J., Jr. (1982) J. Biol. Chem. *257*, 5852-5860.
61. Emr, S.D., Hall, M.N. & Silhavy, T.J. (1980) J. Cell Biol. *86*, 701-711.
62. Emr, S.D., Hanley-Way, S. & Silhavy, T.J. (1981) Cell *23*, 79-88.
63. Ito, K., Wittekind, M., Nomura, M., Shiba, K., Yura, T., Miura, A. & Nashimoto, H. (1983) Cell *32*, 789-797.
64. Shiba, K., Ito, K., Yura, T. & Cerretti, D.P. (1984) EMBO J. *3*, 631-635.
65. Blobel, G., Walter, P., Chang, C.N., Goldman, B.M., Erickson, A.H. & Lingappa, V.R. (1979) In Society for Experimental Biology Symposium, vol. 33, 9-36, Hopkins, C.R. & Duncan, C.J., eds. Cambridge University Press, Great Britain.
66. Blobel, G. (1979) In From Gene to Protein: Information Transfer in Normal and Abnormal Cells, 347-360, Russell, T.R., Brew, K., Faber, H. & Schultz, J., eds. Academic Press, New York.
67. Blobel, G. (1980) Proc. Natl. Acad. Sci. USA *77*, 1496-1500.
68. Moreno, F., Fowler, A.V., Hall, M., Silhavy, T.J., Zabin, I. & Schwartz, M. (1980) Nature *286*, 356-359.
69. Kadonaga, J.T., Gautier, A.E., Straus, D.H., Charles, A.D., Edge, M.D. & Knowles, J.R. (1984) J. Biol. Chem. *259*, 2149-2154.
70. Benson, S.A. & Silhavy, T.J. (1983) Cell *32*, 1325-1335.
71. Russell, M. & Model, P. (1981) Proc. Natl. Acad. Sci. USA *78*, 1717-1721.
72. Koshland, D., Sauer, R.T. & Botstein, D. (1982) Cell *30*, 903-914.
73. Inouye, S., Hsu, C.S., Itakura, K. & Inouye, M. (1983) Science *221*, 59-61.
74. Inouye, S., Vlasuk, G.P., Hsiung, H. & Inouye, M. (1984) J. Biol. Chem. *259*, 3729-3733.
75. Hayashi, S., Chang, S.Y., Chang, S. & Wu, H.C. (1984) J. Biol. Chem. *259*, 10448-10454.
76. Talmadge, K., Kaufman, J. & Gilbert, W. (1980) Proc. Natl. Acad. Sci. USA *77*, 3988-3992.
77. Müller, M., Ibrahimi, I., Chang, C.N., Walter, P. & Blobel, G. (1982) J. Biol. Chem. *257*, 11860-11863.
78. Watts, C., Wickner, W. & Zimmermann, R. (1983) Proc. Natl. Acad. Sci. USA *80*, 2809-2913.

79. Talmadge, K., Brosius, J. & Gilbert, W. (1981) Nature *294*, 176–178.
80. Giam, C.Z., Hayashi, S. & Wu, H.C. (1984) J. Biol. Chem. *259*, 5601–5605.
81. Tokunaga, M., Loranger, J.M., Wolfe, P.B. & Wu, H.C. (1982) J. Biol. Chem. *257*, 9922–9925.
82. Tokunaga, M., Loranger, J.M. & Wu, H.C. (1984) J. Cell Biochem. *24*, 113–120.
83. Nakamura, K. & Inouye, M. (1980) Proc. Natl. Acad. Sci. USA *77*, 1369–1373.
84. Yamagata, H., Nakamura, K. & Inouye, M. (1981) J. Biol. Chem. *256*, 2194–2198.
85. Huang, Y.X., Ching, G. & Inouye, M. (1983) J. Biol. Chem. *258*, 8139–8145.
86. Ogata, R.T., Winters, C. & Levine, R.P. (1982) J. Bacteriol. *151*, 819–827.
87. Lin, J.J.C., Kanazawa, H., Ozols, J. & Wu, H.C. (1978) Proc. Natl. Acad. Sci. USA *75*, 4891–4895.
88. Inouye, S., Franceschini, T., Sato, M., Itakura, K. & Inouye, M. (1983) EMBO J. *2*, 87–91.
89. Ghrayeb, J. & Inouye, M. (1984) J. Biol. Chem. *259*, 463–467.
90. Josefsson, L.G. & Randall, L.L. (1981) Cell *25*, 151–157.
91. Meek, R.L., Walsh, K.A. & Palmiter, R.D. (1982) J. Biol. Chem. *257*, 12245–12251.
92. Hortin, G. & Boime, I. (1981) J. Biol. Chem. *256*, 1491–1494.
93. Lai, J.S., Sarvas, M., Brammar, W.J., Neugebauer, K. & Wu, H.C. (1981) Proc. Natl. Acad. Sci. USA *78*, 3506–3510.
94. Smith, W.P., Tai, P.C. & Davis, B.D. (1981) Proc. Natl. Acad. Sci. USA *78*, 3501–3505.
95. Nielson, J.B.K., Caulfield, M.P. & Lampen, J.O. (1981) Proc. Natl. Acad. Sci. USA *78*, 3511–3515.
96. Nielsen, J.B.K. & Lampen, J.O. (1982) J. Biol. Chem. *257*, 4490–4496.
97. Inukai, M., Takeuchi, M., Shimizu, K. & Arai, M. (1979) J. Bacteriol. *140*, 1098–1101.
98. Hussain, M., Ichihara, S. & Mizushima, S. (1980) J. Biol. Chem. *255*, 3707–3712.
99. Inukai, M. & Inouye, M. (1983) Eur. J. Biochem. *130*, 27–32.
100. Lin, J.J.C., Kanazawa, H. & Wu, H.C. (1980) J. Bacteriol. *141*, 550–557.
101. Silhavy, T.J., Shuman, H.A., Beckwith, J. & Schwartz, M. (1977) Proc. Natl. Acad. Sci. USA *74*, 5411–5415.
102. Jackson, R.C. & Blobel, G. (1980) Annals NY Acad. Sci. *343*, 391–403.
103. Zimmerman, M., Ashe, B.M., Alberts, A.W., Pierzchala, P.A., Powers, J.C., Nishino, N., Strauss, A.W. & Mumford, R.A. (1980) Annals NY Acad. Sci. *343*, 405–414.
104. Mollay, C., Vilas, U. & Kreil, G. (1982) Proc. Natl. Acad. Sci. USA *79*, 2260–2263.
105. Daum, G., Gasser, S.M. & Schatz, G. (1982) J. Biol. Chem. *257*, 13075–13080.
106. Böhni, P.C., Daum, G. & Schatz, G. (1983) J. Biol. Chem. *258*, 4937–4943.
107. Cerletti, N., Böhni, P.C. & Suda, K. (1983) J. Biol. Chem. *258*, 4944–4949.
108. Lively, M.O. & Walsh, K.A. (1983) J. Biol. Chem. *258*, 9488–9495.
109. Lively, M.O. & Walsh, K.A. (1983) In Protein Transport and Secretion, 379–387, Oxender, D., ed. Alan R. Liss, New York.
110. Brennan, M.D., Warren, T.G. & Mahowald, A.P. (1980) J. Cell Biol. *87*, 516–520.
111. Jackson, R.C. & White, W.R. (1981) J. Biol. Chem. *256*, 2545–2550.
112. Zwizinski, C. & Wickner, W. (1980) J. Biol. Chem. *255*, 7973–7977.
113. Date, T. & Wickner, W. (1981) Proc. Natl. Acad. Sci. USA *78*, 6106–6110.
114. Wolfe, P.B., Silver, P. & Wickner, W. (1982) J. Biol. Chem. *257*, 7898–7902.

115. Wolfe, P.B., Wickner, W. & Goodman, J.M. (1983) J. Biol. Chem. *258*, 12073–12080.
116. Zwizinski, C., Date, T. & Wickner, W. (1981) J. Biol. Chem. *256*, 3593–3597.
117. Ohno-Iwashita, Y. & Wickner, W. (1983) J. Biol. Chem. *258*, 1895–1900.
118. Ito, K. (1982) J. Biol. Chem. *257*, 9895–9897.
119. Zimmerman, R. & Wickner, W. (1983) J. Biol. Chem. *258*, 3920–3925.
120. Date, T. (1983) J. Bacteriol. *154*, 76–83.
121. Innis, M.A., Tokunaga, M., Williams, M.E., Loranger, J.M., Chang, S.Y., Chang, S. & Wu, H.C. (1984) Proc. Natl. Acad. Sci. USA *81*, 3708–3712.
122. Inukai, M., Takeuchi, M., Shimuzu, K. & Arai, M. (1978) J. Antibiotics *31*, 1203–1205.
123. Dev, I.K. & Ray, P.H. (1984) J. Biol. Chem. *259*, 11114–11120.
124. Tokunaga, M., Loranger, J.M. & Wu, H.C. (1984) J. Biol. Chem. *259*, 3825–3830.
125. Yamada, H., Yamagata, H. & Mizushima, S. (1984) FEBS Lett. *166*, 179–182.
126. Tokunaga, H. & Wu, H.C. (1984) J. Biol. Chem. *259*, 6098–6104.
127. Tokunaga, M., Loranger, J.M. & Wu, H.C. (1983) J. Biol. Chem. *258*, 12102–12105.
128. Yamagata H., Daishima, K. & Mizushima, S. (1983) FEBS Lett. *158*, 301–304.
129. Yamagata, H., Ippolito, C., Inukai, M. & Inouye, M. (1982) J. Bacteriol. *152*, 1163–1168.
130. Regue, M., Remenick, J., Tokunaga, M., Mackie, G.A. & Wu, H.C. (1984) J. Bacteriol. *158*, 632–635.
131. Yamagata, H., Taguchi, N., Daishima, K. & Mizushima, S. (1983) Mol. Gen. Genet. *192*, 10–14.
132. Silver, P. & Wickner, W. (1983) J. Bacteriol. *154*, 569–572.
133. Date, T., Zwizinski, C., Ludmerer, S. & Wickner, W. (1980) Proc. Natl. Acad. Sci. USA *77*, 827–831.
134. Daniels, C.J., Bole, D.G., Quay, S.C. & Oxender, D. (1981) Proc. Natl. Acad. Sci. USA *78*, 5396–5400.
135. Enequist, H.G., Hirst, T.R., Harayama, S., Hardy, S.J.S. & Randall, L.L. (1981) Eur. J. Biochem. *116*, 227–233.
136. Horiuchi, S., Marty-Mazars, D., Tai, P.C. & Davis, B.D. (1983) J. Bacteriol. *154*, 1215–1221.
137. Marty-Mazars, D., Horiuchi, S., Tai, P.C. & Davis, B.D. (1983) J. Bacteriol. *154*, 1381–1388.
138. Horiuchi, S., Tai, P.C. & Davis, B.D. (1983) Proc. Natl. Acad. Sci. USA *80*, 3287–3291.
139. Oliver, D.B. & Beckwith, J. (1982) Cell *30*, 311–319.
140. Tsai, H., Kula, M.R. & Lindenmaier, W. (1982) In Genetic Engineering Techniques: Recent Developments, 159–172, Huang, P.C., Kuo, T.T. & Wu, R., eds. Academic Press, New York.

4

The Posttranslational Processing of the Precursors of Secreted Peptides

Günther Kreil

Secreted polypeptides pass through a complex intracellular route before reaching the cell surface, and this route is used, to some extent, by plasma membrane proteins as well. Among the former, secreted peptides are of particular interest, as they originate from larger precursors from which they are liberated by different processing reactions in the course of this intracellular transport. In common with other secreted polypeptides, these precursors contain as nascent chains a prepeptide or signal peptide that is cleaved early, probably cotranslationally, as demonstrated for larger polypeptides. The resulting propeptides are then transported to the Golgi apparatus, where they apparently enter a special route by being packaged into secretory granules. Most, and possibly all, of the further processing takes place in these granules. The mature peptides are then stored and finally released in response to a variety of physiological stimuli. In recent years, it has been demonstrated that several biologically active peptides of diverse functions may be liberated from a given precursor. Morover, different processing routes may operate in different cell types. In spite of this complex picture, it still appears that only a few types of enzymatic reactions are involved in the liberation of most peptides serving as hormones, neurotransmitters, or in defensive secretions. Our current knowledge about the enzymes catalyzing these typical cleavages is summarized in this review. The maturation of prohormones has recently been reviewed by Docherty and Steiner (1), and I will thus concentrate on the findings made during the last years.

Precursors and Their Initial Processing by Endoproteases

Sequence analysis of porcine (2) and bovine (3) proinsulin revealed the general structure (B-chain)-arg-arg-(C-peptide)-lys-arg(A-chain) for this precursor. The pairs of lysine/arginine residues marking the boundaries between the proregions and the final product turned out to be a characteristic feature of a large number of peptide precursors. This arrangement was subsequently detected in the precursors of parathyroid hormone (4), gastrin (5), somatostatin (6), and a few others. With the advent of recombinant DNA technology, the structure of many more precursors of physiologically active peptides was determined by cloning and sequencing the corresponding cDNAs. These studies also revealed a new level of complexity, when it became clear that several peptides of similar or diverse structure and function could be derived from one precursor. This was first demonstrated for the common precursor of corticotropin, β-lipotropin, and several of its active fragments (7). In this and other cases, the different end-products are flanked by pairs of basic residues, which further emphasize the importance of this processing signal. It is noteworthy that such "polyproteins" have recently been found in invertebrates (8) and even in yeast (9); and again, initial processing sites are marked by lysine/arginine pairs. Thus, the liberation of biologically active fragments from larger precursors via cleavage at pairs of basic amino acids is a very old mechanism that apparently dates back to the early phases of the evolution of eukaryotes.

Table 4.1 lists a number of precursors where cleavage at lysine/arginine sequences must take place to yield the mature peptides. These are divided into "simple" precursors that yield only one product and "complex" ones, where more than one molecule of a peptide with a biological function is generated during processing. This division is somewhat arbitrary, since even processing of simple precursors can proceed via distinct intermediates that already have some of the physiologic activity of the final product. This is exemplified by the coexistence and simultaneous release of gastrin-34 and gastrin-17 (10). Moreover, the discovery that a single polypeptide chain can give rise to several active fragments has led to the general notion that putative prosegments should be tested for possible functions. Unknown sequences flanked by pairs of basic residues are particularly promising candidates for such tests.

In spite of numerous efforts in different laboratories, the enzyme(s) catalyzing this processing reaction at pairs of basic amino acids has yet to be fully characterized. Granule preparations from anglerfish islets have been shown to convert proinsulin and proglucagon to mature products (43). The pH optimum of this conversion was around 5, and activity was present in the soluble as well as in the membrane fraction of these granules. Antipain, leupeptin, and p-chloromercuribenzoate inhibited the processing reaction, while inhibitors of serine proteases had no effect. A similar enzymatic activity has been shown to be present in lysates of secretory granules from rat pituitary (44). The activity profile against pro-opiomelanocortin showed some differences for granules isolated from different parts of the pituitary (45). Recently, isolation of the protease that converts proinsulin to insulin from rat islets of Langerhans has been attempted. With the active site-directed reagent (^{125}I)Tyr-Ala-Lys-Arg•CH$_2$Cl, it was found that in granule preparations, a protein with a molcular weight of about 32,000 was mainly labeled by this inhibitor (46). However, it was subsequently shown that

Table 4.1. Precursors of Physiologically Active Peptides

Simple precursors

Prepro
  Insulin (11–15)
  Parathyroid hormone (16)
  Somatostatin (17–20)
  Relaxin (21)
  Gastrin (22, 23)
  Corticotropin-releasing factor (24)
  Growth hormone-releasing factor (25, 26)
  PYL[a] (27)

Complex precursors (polyproteins)	Products
Prepro	
Opiomelanocortin (7)	ACTH; α-, β-, and γ-MSH; β- and γ-lipotropin; β-endorphin
Enkephalin A (28)	Met-Enkephalin (4 copies), Leu-enkephalin, Met-enk.-Arg-Phe, Met-enk.-Arg-Gly-Leu
Enkephalin B (29)	β-Neoendorphin, dynorphin, Leu-enkephalin with COOH terminal extension
Vasopressin (30)	Vasopressin, neurophysin II
Oxtocin (31)	Oxytocin, neurophysin I
Glucagon (32–34)	Glucagon, 1 or 2 glucagon-related peptides
Calcitonin (35–38)	Calcitonin and PDN-21 (thyroid) calcitonin-gene-related peptide (brain)
VIP (39)	Vasoactive intestinal Peptide, PHM-27
Substance P (40)	Substance P, substance K
Caerulein (41)	Caerulein (1–3 copies)
TRH (42)	Thyrotropin-releasing hormone (4 copies)
Egg-laying hormone (8)	Egg-laying hormone, peptides A and B
α-mating factor (9, 76)	α-mating factor (2 or 4 copies)

References are indicated in parentheses.
[a] Amide.

this protein could be precipitated by antibodies against cathepsin B (47), which suggests that it originates from lysosomes present in the granule preparations.

Present evidence then shows that the protease cleaving after pairs of basic amino acids has an acidic pH optimum and is susceptible to SH reagents. Its distinction from cathepsin B rests on somewhat different inhibition patterns with a variety of SH reagents. Moreover, cathepsin B appears to have too broad a substrate specificity to catalyze the very limited proteolysis that takes place during activation of these precursors.

The cleavage of hormone precursors at pairs of lysine/arginine residues is probably a late reaction in the secretory pathway. Experiments by Kelly and coworkers have shown that in AtT-20 cells, a mouse pituitary cell line, both a constitutive and a regulated secretory pathway exist (48–50). The cleavage products of pro-opiomelanocortin are transported via the latter pathway, in that they are stored in secretory granules and only released in the presence of a secretagogue. On the other hand, a membrane glycoprotein encoded in this cell line by an endogenous type C virus enters the constitutive pathway. In the presence of chloroquine, which is known to raise the pH in intracellular compartments, increasing amounts of intact precursor were released via the constitutive pathway ([51], and discussion in ref [52]). More recently, it has been demonstrated that in AtT-20 cells transformed with an SV40 -pBR322 recombinant vector containing the cDNA for human preproinsulin, insulin is generated, stored, and released upon stimulation with a secretagogue (53). On the other hand, fibroblast L cells transformed with this vector secrete only proinsulin in a constitutive way. This work clearly suggests that prohormones enter a separate, regulated pathway of secretion and, in keeping with the *in vitro* results, that processing occurs late in secretory granules in an acidic environment.

The prohormone-processing enzymes are not present in all cells. Several groups have shown that fibroblasts transfected with the insulin gene secrete proinsulin, but not insulin (53–55). Also, frog oocytes, which have been used extensively to study the export of proteins after injecting the respective mRNAs, do not liberate insulin (56) or vasopressin (57) from their precursors.

In several of the precursors mentioned in Table 4.1, release of the final product(s) must also involve processing at single basic, mostly arginine residues. Examples are the precursors for somatostatin (formation of somatostatin-28, [17–20]), vasopressin-neurophysin II (cleavage of the COOH terminal glycopeptide [30]), and vasoactive intestinal peptide (VIP) (39). Cleavage at single basic amino acids is the hallmark of extracellular zymogen activation, as exemplified by the activation of blood-clotting factors. It is currently not known whether the intracellular processing at single arginine residues is catalyzed by the same enzyme(s) that hydrolyzes at pairs of basic amino acids. In frog oocytes, the vasopressin-neurophysin precursor is apparently cleaved only at the single arginine separating the terminal glycopeptide from neurophysin II (57), but not at the Lys-Arg sequence separating vasopressin from the latter. This might then indicate that different proteases catalyze these two processing reactions.

Finally, it should be mentioned that processing of several precursors must also involve endoproteolytic cleavages at sites not marked by the presence of basic amino acids (see [1] for a review). A chymotrypsin-like enzyme has been isolated from pig pituitary granules (58), but overall, very little is known about the proteases catalyzing these atypical processing reactions.

Two interesting, but exceptional, cases are the precursors of nerve growth factor (NGF) and epidermal growth factor (EGF), which are both synthesized in rather high amounts in the submaxillary gland of male mice. The precursors of NGF (59, 60) and EGF (61) have both been analyzed via cloning of the corresponding cDNAs. NGF is secreted as part of a 7S complex that also contains an arginine-specific esterase probably involved in the processing of NGF precursor. By far, the most complex precursor found to date is the one for EGF, which is a polypeptide comprised of 1217 amino acids and is thus 5 to 10 times larger than typical precursors for physiologically active peptides. It contains a single copy of EGF, a 58 amino acid peptide, and seven peptides related to different degrees to EGF. Putative processing of this precursor to yield these different fragments would involve cleavages at both single and pairs of basic amino acids, as well as at other types of proteolysis. The reason for the unusual size and complexity of the EGF precursor is currently not understood.

Enzymatic Reactions at the Termini of Processing Intermediates

Cleavage of Basic Amino Acids

After the endoproteolytic hydrolysis of precursors, the lysine and arginine residues present at the newly formed carboxyl ends have to be cleaved by a carboxypeptidase B-like enzyme. Such an enzyme has been described in crude granule preparations from rat islets (62), and has recently been characterized in more detail using granules from a transplantable rat insulinoma as starting material (63). It is a soluble carboxypeptidase with an acidic pH optimum and a molecular weight of about 54,000. Highest activity was detected in the presence of cobalt ions. A similar enzyme had been isolated earlier from a granule fraction of bovine adrenal medulla (64, 65), which apparently participates in the liberation of enkephalins from proenkephalin A. This enzyme, termed enkephalin convertase, has now been isolated from bovine brain and pituitary as well, and has been purified to homogeneity (66). Available evidence suggests that identical carboxypeptidases with molecular weights of about 50,000 are present in the bovine brain, pituitary, and adrenal medulla. Like the rat islet enzyme, enkephalin convertase is stimulated by cobalt ions and inhibited by p-chloromercuriphenylsulphonate and 1,10-phenanthroline. It is quite likely that bovine enkephalin convertase and the carboxypeptidase isolated from rat islets are homologous enzymes that catalyze the cleavage of basic amino acids from the intermediates generated during the processing of prohormones in a variety of cells.

Formation of Terminal Amides

Many peptide hormones and other physiologically active peptides terminate not with a free α-carboxyl group, but instead with an amide. The distribution and biosynthesis of these terminal amides have recently been reviewed (67). Studies of the biosynthesis of melittin, the main constituent of bee venom, have demonstrated that in the precursor, a glycine residue is present adjacent to the carboxy terminal amino acids of the end-product (68). Glycine residues have since been found at the same position in the precursors of a variety of amidated peptides, including calcitonin (35), gastrin (22), vasopressin (30), oxytocin (31), corticotropin-releasing factor (24), substance P (40), VIP (39), caerulein (41), egg-laying hormone (8), and thyrotropin-releasing hormone (42). The formation of terminal amides proceeds by oxidative cleavage of the terminal glycine, whereby its nitrogen remains in the peptide (69). The enzyme-catalyzing this reaction has been partially purified from mammalian pituitary. Besides molecular oxygen, copper ions and ascorbic acid are apparently required for this redox reaction (70). Peptides with a terminal D-alanine instead of glycine are also converted to the corresponding amides (71). It is puzzling that contrary to the other processing reactions, the amidation has a pH optimum of about 7.

Stepwise Cleavage of Dipeptides From the Amino Terminus

Studies on the biosynthesis of melittin led to the discovery of a new type of processing reaction that proceeds via stepwise cleavage of dipeptides. The proregion of promelittin was found to have a very unusual sequence in that even numbered residues were all proline or alanine (72). Thus, the liberation of melittin from this precursor must differ from typical prohormone-hormone conversions, since ultimately an alanyl-glycine bond has to be hydrolyzed. In the search for the relevant enzymes in extracts from honeybee venom glands, high activities of a dipeptidylaminopeptidase were detected with ala-pro- and ala-ala-p-nitroanilides as substrates (73). It could be subsequently shown that this enzyme cleaves the proregion in a stepwise fashion, starting from the amino end (74). The dipeptides thus liberated were further hydrolyzed to free amino acids by a dipeptidase. The specificity of this sequential activation of a precursor appears to be solely governed by the nature of the second amino acid, which has to be either proline, alanine, or possibly glycine. The activation of promelittin is probably an extracellular reaction as the bulk of the dipeptideylaminopeptidase is present in a soluble form in the venom sac. Such a late reaction is not surprising in view of the fact that melittin has strong lytic properties, and therefore must be liberated only in an environment where it can no longer interact with phospholipid bilayers; *i.e.*, in the venom sac, which as a strong chitin wall.

The same type of stepwise activation has recently been demonstrated for the precursor of yeast α-mating factor (75). In *Saccharomyces cerevisiae*, two genes for this precursor have been identified that contain two or four copies of the mature peptide (9, 76). These copies are separated by so-called spacer peptides of the general structure Lys-Arg-(X-Ala)$_{2-3}$, with X being mostly Glu or Asp; in one instance each, also Asn and Val. In other *Saccharomyces* species, variants of this precursor with three to five copies of the α-mating factor have been detected (77). The typical processing of this yeast precursor by cleavage at, and excision of, the pair of basic amino acids would yield an intermediate form of α-mating factor with an amino terminal extension. It has been shown by Julius and coworkers (75) that the yeast mutant *ste*13 does indeed secrete such intermediates, which have a very low biological activity. This mutant lacks a membrane-bound dipeptidylaminopeptidase, which is required for the liberation of the final product. The defect could, in fact, be corrected by introducing the cloned gene for this enzyme into the mutant.

A third example of this type of processing has apparently been detected in frog skin. Via cDNA cloning, the sequence of the precursor for caerulein, a peptide related to mammalian cholecystokinin and gastrin, could be determined (41). In the different variants of this precursor, a larger form of caerulein with the amino terminal extension Phe-Ala-Asp-Gly is flanked by pairs of basic residues. This led to the assumption that a dipeptidylaminopeptidase may also be involved in the processing of caerulein precursors. Skin secretions of *Xenopus laevis* do indeed contain high dipeptidylaminopeptidase activity (A. Hutticher and G. Kreil, in preparation), while aminopeptidases are barely detectable. From the recently determined sequence of the precursor of xenopsin (78) (another peptide secreted by skin of *X. laevis*), the action of dipeptidylpeptidase can also be invoked in the liberation of this peptide. It thus appears quite likely that for a variety of precursors, sequential cleavage of dipeptides is a late step in the processing cascade. Due to its intricate specificity, this reaction can take place in the presence of many other polypeptides, both in- and outside the cell.

Conclusions

The biosynthesis of secreted peptides functioning as hormones, neurotransmitters, mating factors, repellents, toxins, etc. is clearly of some interest. It appears likely that many of these are transported via a specific route involving storage in intracellular granules. The export of these peptides is often not only regulated at the level of their synthesis, but even more so at the rate of their release. In this respect, they differ from many secreted proteins found in blood, an extracellular matrix, etc., as well as from plasma membrane proteins.

The formation of such peptides via precursors is an old process dating back to the early phases of eukaryotic evolution. Small peptides may have played a major role in early types of cell-cell interactions and in the evolution of the nervous system. Current evidence suggests that a polypeptide must have a minimum size to be recognized by signal recognition particle (SRP) and then threaded through the membrane of the endoplasmic reticulum (ER). The smallest prepropeptide known to date contains 64 amino acids (27), and this may well be close to the minimum. Any smaller peptide must then originate from a larger precursor, and limited, specific proteolysis becomes essential. With the advent of recombinant DNA technology, the analysis of the structure of such precursors via cDNA cloning has become much easier. The sequences of many precursors for mammalian hormones, etc. are now known, and some invertebrate sequences have also been elucidated. The methods currently available will also permit the search for peptides and their precursors in unicellular eukaryotic and prokaryotic cells, as well as in plants. Knowledge of the structure of precursors then opens the way towards investigations of the processing mechanisms and other aspects of the intracellular transport that play a role in the biosynthesis of peptides of diverse functions.

References

1. Docherty, K. & Steiner, D.F. (1982) Ann. Rev. Physiol. *44*, 625-638.
2. Chance, R.E., Ellis, R.M. & Bromer, W.W. (1968) Science *161*, 165-167.
3. Nolan, C., Margoliash, E., Peterson, J.D. & Steiner, D.F. (1971) J. Biol. Chem. *246*, 2780-2795.
4. Hamilton, J.W., Niall, H.D., Jacobs, J.W., Keutmann, H.T., Potts, J.T., Jr. & Cohn, D.V. (1974) Proc. Natl. Acad. Sci. USA *71*, 653-656.
5. Gregory, R. (1974) J. Physiol. *241*, 1-31.
6. Pradayrol, L., Jörnvall, H., Mutt, V. & Ribet, A. (1980) FEBS Lett. *109*, 55-58.
7. Nakanishi, S., Inoue, A., Kita, T., Nakamura, M., Chang, A.C.Y., Cohen, S.N. & Numa, S. (1979) Nature *278*, 423-427.
8. Scheller, R.H., Jackson, J.F., McAllister, L.B., Rothman, B.S., Mayer, E. & Axel, R. (1983) Cell *32*, 7-22.
9. Kurjan, J. & Herskowitz, I. (1982) Cell *30*, 933-943.
10. Dockray, G.J., Vaillant, C. & Hopkins, C.R. (1978) Nature *273*, 770-772.
11. Chan, S.J., Kaim, P. & Steiner, D.F. (1976) Proc. Natl. Acad. Sci. USA *73*, 1964-1968.
12. Villa-Komaroff, L., Efstratiadis, A., Broome, S., Lomedico, P., Tizard, R., Naber, S.P., Chick, W.L. & Gilbert, W. (1978) Proc. Natl. Acad. Sci. USA *75*, 3727-3731.
13. Ullrich, A., Dull, T.J., Gray, A., Brosius, J. & Sures, I. (1980) Science *209*, 612-615.
14. Hobart, P.M., Shen, L., Crawford, R., Pictet, R.L. & Rutter, W.J. (1980) Science *210*, 1360-1363.
15. Cahn, S.J., Emdin, S.O., Kwok, S.C.M., Kramer, J.M., Falkmer, S. & Steiner, D.F. (1981) J. Biol. Chem. *256*, 7595-7602.
16. Habener, J.F., Rosenblatt, M., Kemper, B., Kronenberg, H.M., Rich, A. & Potts, J.R., Jr. (1978) Proc. Natl. Acad. Sci. USA *75*, 2616-2620.
17. Hobart, P., Crawford, R., Shen, L., Pictet, R. & Rutter, W.J. (1980) Nature (London) *288*, 137-141.
18. Goodman, R.H., Jacobs, J.W., Chin, W.W., Lund, P.K., Dee, P.C. & Haabener, J.F. (1980) Proc. Natl. Acad. Sci. USA *77*, 5869-5873.
19. Goodman, R.H., Jacobs, J.W., Dee, P.C. & Habener, J.F. (1982) J. Biol. Chem. *257*, 1156-1159.
20. Minth, C.D., Taylor, W.L., Magazin, M., Tavianini, M.A., Collier, K., Weith, H.L. & Dixon, J.E. (1982) J. Biol. Chem. *257*, 10372-10377.
21. Hudson, P., Haley, J., Cronk, M., Shine, J. & Niall, H. (1981) Nature *291*, 127-131.
22. Yoo, O.J., Powell, C.T. & Agarwal, K.L. (1982) Proc. Natl. Acad. Sci. USA *79*, 1049-1053.
23. Boel, E., Vuust, J., Norris, F., Norris, K., Wind, A., Rehfeld, F.F. & Marcker, K.A. (1983) Proc. Natl. Acad. Sci. USA *80*, 2866-2869.
24. Furutani, Y., Morimoto, Y., Shibahara, S., Noda, M., Takahashi, H., Hirose, T., Asai, M., Inayama, S., Hayashida, H., Miyata, T. & Numa, S. (1983) Nature *301*, 537-540.

25. Gubler, U., Monahan, J.J., Lomedico, P.T., Bhatt, R.S., Collier, K.J., Hoffmann, B.J., Böhlen, P., Esch, F., Ling, N., Zeytin, F., Brazeau, P., Poonian, M.S. & Gage, L.P. (1983) Proc. Natl. Acad. Sci. USA 80, 4311–4314.
26. Mayo, K., Vale, W., Rivier, J., Rosenfeld, M.G. & Evans, R.M. (1983) Nature 306, 86–88.
27. Hoffmann, W., Richter, K. & Kreil, G. (1983) EMBO J. 2, 711–714.
28. Noda, M., Furutani, Y., Takahashi, H., Toyosato, M., Hirose, T., Inayama, S., Nakanishi, S. & Numa, S. (1982) Nature 295, 202–206.
29. Kakidani, H., Furutani, Y., Takahashi, H., Noda, M., Morimoto, Y., Hirose, T., Asai, M., Inayama, S., Nakanishi, S. & Numa, S. (1982) Nature 298, 245–249.
30. Land, H., Schütz, G., Schmale, H. & Richter, D. (1982) Nature 295, 299–303.
31. Land, H., Grez, M., Ruppert, S., Schmale, H., Rehbein, M., Richter, D. & Schütz, G. (1983) Nature 302, 342–344.
32. Lund, P.K., Goodman, R.H., Montminy, M.R., Dee, P.C. & Habener, J.K. (1983) J. Biol. Chem. 258, 3280–3284.
33. Bell, G.I., Santerre, R.F. & Mullenbach, G.T. (1983) Nature 302, 716–718.
34. Lopez, L.C., Frazier, M.L., Su, C.-J., Kumar, A. & Saunders, G.F. (1983) Proc. Natl. Acad. Sci. USA 80, 5485–5489.
35. Amara, S.G., David, D.N., Rosenfeld, M.G., Ross, B.A. & Evans, R.M. (1980) Proc. Natl. Acad. Sci. USA 77, 4444–4448.
36. Jacobs, J.W., Goodman, R.H., Chin, W.W., Dee, P.C., Habener, J.F., Bell, N.H. & Potts, J.T., Jr. (1981) Science 213, 457–459.
37. MacIntyre, I., Hillyard, C.J., Murphy, P.K., Reynolds, J.J., Gaines Das, R.E. & Craig, R.K. (1982) Nature 300, 460–463.
38. Amara, S.G., Jonas, V., Rosenfeld, M.G., Ong, E.S. & Evans, R.M. (1982) Nature 298, 240–244.
39. Itoh, N., Obata, K., Yanaihara, N. & Okamoto, H. (1983) Nature 304, 547–550.
40. Nawa, H., Hirose, T., Takashima, H., Inayama, S. & Nakanishi, S. (1983) Nature 306, 32–36.
41. Hoffmann, W., Bach, T.C., Seliger, H. & Kreil, G. (1983) EMBO J. 2, 111–114.
42. Richter, K., Kawashima, E., Egger, R. & Kreil, G. (1984) EMBO J. 3, 617–621.
43. Fletcher, D.J., Quigley, J.P., Bauer, G.E. & Noe, B.D. (1981) J. Cell Biol. 90, 312–322.
44. Loh, Y.P. & Gainer, H. (1982) Proc. Natl. Acad. Sci. USA 79, 108–112.
45. Loh, Y.P. & Chang, T.-L. (1982) FEBS Lett. 137, 57–62.
46. Docherty, K., Carroll, R.J. & Steiner, D.F. (1982) Proc. Natl. Acad. Sci. USA 79, 4613–4617.
47. Docherty, K., Carroll, R. & Steiner, D.F. (1983) Proc. Natl. Acad. Sci. USA 80, 3245–3249.
48. Gumbiner, B. & Kelly, R.B. (1981) Proc. Natl. Acad. Sci. USA 78, 318–322.
49. Moore, H.P., Gumbiner, B. & Kelly, R.B. (1983) Nature 302, 434–436.
50. Gumbiner, B. & Kelly, R.B. (1982) Cell 28, 51–59.
51. Moore, H.-P., Gumbiner, B. & Kelly, R.B. (1983) Nature 302, 434–436.
52. Dean, R.T. (1983) Nature 305, 73–74 (and reply by Moore, et al.).
53. Moore, H.-P., Walker, M.D., Lee, F. & Kelly, R.B. (1983) Cell 35, 531–538.
54. Gruss, P. & Khoury, G. (1981) Proc. Natl. Acad. Sci. USA 78, 133–137.
55. Lomedico, P.T. (1982) Proc. Natl. Acad. Sci. USA 79, 5798–5802.
56. Rapoport, T.A., Thiele, B.J., Behn, S., Marbaix, G., Cleuter, Y., Hubert, E. & Huez, G. (1978) Eur. J. Biochem. 87, 229–233.

57. Richter, D. (1983) Trends Biochem. Sci. *8*, 278–281.
58. Seidah, N.G., Pelaprat, D., Lazure, C. & Chretien, M. (1983) FEBS Lett. *159*, 68–72.
59. Scott, J., Selby, M., Urdea, M., Quiroga, M., Bell, G.I. & Rutter, W.J. (1983) Nature *302*, 538–540.
60. Ullrich, A., Gray, A., Berman, C. & Dull, T.J. (1983) Nature *303*, 821–825.
61. Scott, J., Urdea, M., Quiroga, M., Sanchez-Pescador, R., Fong, N., Selby, M., Rutter, W.J. & Bell, G.I. (1983) Science *221*, 236–240.
62. Kemmler, W., Steiner, D.F. & Borg, J. (1973) J. Biol. Chem. *248*, 4544–4511.
63. Docherty, K. & Hutton, J.C. (1983) FEBS Lett. *162*, 137–141.
64. Hook, V.Y.H., Eiden, L.E. & Brownstein, M.J. (1982) Nature *295*, 341–342.
65. Fricker, L.D. & Snyder, S.H. (1982) Proc. Natl. Acad. Sci. USA *79*, 3886–3890.
66. Fricker, L.D. & Snyder, S.H. (1983) J. Biol. Chem. *258*, 10950–10955.
67. Kreil, G. (1984) Methods in Enzymol. vol. *106*, 218–223.
68. Suchanek, G. & Kreil, G. (1977) Proc. Natl. Acad. Sci. USA *74*, 975–978.
69. Bradbury, A.F., Finnie, M.D.A. & Smyth, D.G. (1982) Nature *298*, 686–688.
70. Eipper, B.A., Mains, R.E. & Glembotski, C.C. (1983) Proc. Natl. Acad. Sci. USA *80*, 5144–5148.
71. Landymore-Lim, A.E.N., Bradbury, A.F. & Smyth, D.G. (1983) Biochem. Biophys. Res. Commun. *117*, 289–293.
72. Suchanek, G., Kreil, G. & Hermodson, M.A. (1978) Proc. Natl. Acad. Sci. USA *75*, 701–704.
73. Kreil, G., Mollay, C., Kaschnitz, R., Haiml, L. & Vilas, U. (1980) Ann NY Acad. Sci. *343*, 338–346.
74. Kreil, G., Haiml, L. & Suchanek, G. (1980) Eur. J. Biochem. *111*, 49–58.
75. Julius, D., Blair, L., Brake, A., Sprague, G. & Thorner, J. (1983) Cell *32*, 839–852.
76. Singh, A., Chen, E.Y., Lugovoy, J.M., Chang, C.N., Hitzeman, R.A. & Seeburg, P.H. (1983) Nucl. Acid Res. *11*, 4049–4063.
77. Brake, A.J., Julius, D. & Thorner, J. (1983) Mol. Cell. Biochem. *3*, 1440–1450.
78. Sures, I. & Crippa, M. (1984) Proc. Natl. Acad. Sci. USA *81*, 380–384.

5

Assembly of Multisubunit Membrane Proteins

BARRY E. CARLIN and JOHN PAUL MERLIE

Secretory proteins, lysosomal proteins, and many integral membrane proteins are cotranslationally inserted into the membranes of the rough endoplasmic reticulum (RER) (1). It is now widely held that from the RER, these protein precursors follow a common path to the Golgi, where their paths to subsequent destinations diverge (2). The mechanisms by which these translocations occur are under intense study with regard to many different proteins.

Consideration of the unique constraints imposed by the translocation mechanisms on the assembly of multisubunit membrane proteins leads to some interesting questions: 1) Is assembly an efficient process? If the process of subunit association is stochastic, it would seem difficult for polypeptide chains synthesized in different regions of the endoplasmic reticulum (ER) to be efficiently and rapidly assembled. If assembly is not simply dependent upon the relative frequency of subunit collisions (*i.e.*, if assembly is an active process), the possibility arises that it may be regulated. 2) What is the topology of multisubunit assembly? In what subcellular compartments do subunits assemble, and with what time course with respect to covalent modifications such as protein cleavage, glycosylation, and oligosaccharide-trimming reactions? 3) How does assembly affect or contribute to the ultimate structure (secondary, tertiary, and quaternary) of the individual subunits and, conversely, how does protein structure affect the processes of assembly and transport?

The number of multisubunit membrane proteins that have been characterized in detail, with respect to synthesis, assembly, and intracellular transport, is small. Therefore, we have chosen to review four different examples that appear to be the most thoroughly studied.

IgM

The IgM molecule has been the subject of intensive study due to its critical function in the immune system. IgM not only constitutes 5% of the circulating antibodies, but, in its membrane-bound form, functions as the receptor for antigen-stimulated differentiation of resting B cells to secreting plasma cells. Investigations of the structure, synthesis, and regulation of the protein have benefitted from the availability of cell lines that mimic stages of B cell development, some of which can devote up to 5% of their total synthetic capacity to IgM production. The study of this multisubunit protein has brought to light many interesting features of the assembly process.

Two types of IgM have been distinguished: the membrane-bound form (mIgM) (Figure 5.1) and the secreted form (sIgM) (3). Both forms are composed of covalently linked heavy (μ) and light (L) chains (4–6). A monomer unit with a stoichiometry of μ_2L_2 has been identified in which the μ chains are linked to each other by disulfide linkage in the Fc region (7). In addition, each μ chain is linked to one L chain in the Fab region. sIgM is formed by polymerization of five monomer units with a linking polypeptide J to give a $(\mu_2L_2)_5J$ complex (8) This 900,000 dalton complex is the circulating form of IgM antibody. mIgM remains as a monomer (molecular weight $\cong$ 190,000) anchored in the plasma membrane, where it serves its receptor function.

sIgM and mIgM differ in the structure of their μ chains. Although μm and μs are identical over most of their length, they differ at their C termini (9). Amino acid sequence, deduced from cDNA, demonstrates that μs chains terminate in a 20-amino acid stretch that is hydrophilic in nature, while μm terminates with a 41-amino acid hydrophobic sequence (10, 11). The two chains are translated from distinct mRNAs that arise from alternate splicing of transcripts from a single gene (12). The differing C termini of the μ chains give each type of IgM its unique character. In mIgM, it is the hydrophobic tail sequence of μm that anchors the molecule in the membrane. In sIgM, the hydrophilic tail of μs provides a site

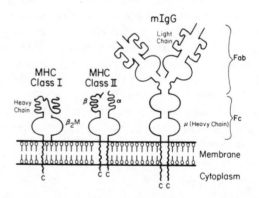

Fig. 5.1. Structures of classes I and II transplantation antigens and mIgM.

for both an additional oligosaccharide linkage and an additional cys residue used in polymerization of the monomer units with each other and the J chain.

Despite the differences in μ chain, the synthesis of the sIgM and mIgM follow very similar pathways. The peptide chains are synthesized on membrane-bound polysomes and undergo cotranslational cleavage of a signal peptide (13). Both types of μ chain undergo cotranslational glycosylation; four sites in μm and five sites in μs (14, 15). Finally, both μs and μm chains bound for externalization transit the Golgi, where terminal sugars are added (16).

The details of the assembly process as they relate to the events described above are best known for the secretory form of IgM, but the general features are shared with mIgM as well as with other Ig molecules (16, 17). In plasma cells, no free μ chain can be detected even after a 3-minute pulse labeling with radioactive amino acids, so that assembly with L chains is quantitative and must occur with nascent or newly synthesized μ chains (18, 19). Radioactive intermediates of the form μL and μ_2L can be detected immediately after such a brief pulse labeling, but within 5 minutes of μ chain completion, nearly all μ is found as μ_2L_2 monomer. L chains, on the other hand, are synthesized in excess and are available for assembly even 30 minutes after translation. In addition, some free L chain is secreted (19).

Polymerization of monomers to pentamers also proceeds at a rapid rate, with half the monomers assembled to pentamers within 15 to 20 minutes of synthesis. Cell fractionation shows that pentamerization has occurred before the μ chains leave the rough microsomal fraction (19).

Glycosylation of μ chains does not appear to have a direct role either in the assembly of μ with L chains or in the polymerization of monomers to pentamers. Inhibition of core glycosylation with tunicamycin does not inhibit assembly of μL complexes in secreting cells, nor is secretion of pentamer prevented (19, 20). In addition, of several cell lines producing mutant μ chains that are deficient in one or more oligosaccharide chains, none are assembly-deficient (21). The effect of tunicamycin on mIgM is less clear. A nonsecreting human cell line has been identified in which underglycosylated polypeptides do not assemble (22); while in other murine lines, assembly is uneffected by tunicamycin (23).

Carboxylic ionophores such as monensin and nigericin are known to disrupt intracellular transport of membrane and secreted proteins and to block addition of terminal sugars to their oligosaccharides (24, 25). Since assembly is complete before IgM chains leave the rough microsomal fraction and since addition of terminal sugars occurs in late Golgi just prior to externalization, it is not surprising that these drugs—though effective in blocking addition of terminal sugars—do not affect chain assembly (19, 20, 26). It is interesting to note that the drugs do inhibit rates of secretion, but that turnover rates of the incompletely processed proteins are no different from those that are terminally processed.

In the process of differentiation of lymphoid stem cells to terminally differentiated secreting plasma cells, several stages can be identified. The first committed cell type to be identified is the pre-B cell (3). Pre-B cells, found in fetal liver and adult bone marrow, are the first cells of the B lineage to synthesize Ig chains, and these are of the μ type (27). More recently, it has been demonstrated that pre-B cells have mRNA for both μs and μm, and that both messages are translated. The μm and μs chains that are produced in these cells undergo signal peptide cleavage and are core glycosylated, but they do not undergo terminal processing. No L chains are synthesized and, hence, no assembly to μ_2L_2 monomers can occur. Finally, little or none of the newly synthesized μ chains are secreted or expressed on the cell surface (see below).

The next stage in the B cell lineage, the resting B cell, can be distinguished from pre-B cells by the expression of mIgM on the cell surface. Cell lines believed to represent late pre-B cells can be stimulated to express mIgM by treatment with bacterial lipopolysaccharide (LPS). The onset of mIgM expression, both in natural and LPS-stimulated differentiation, is tightly coregulated with the induction of L chain synthesis (27, 28). This finding has led to the hypothesis that L chain synthesis followed by assembly of μ and L chains is sufficient and necessary to allow μ chain expression on the cell surface.

Support for this hypothesis comes from studies in which cells having the pre-B cell phenotype (synthesize μ chains, but not L chains) were fused with nonsecreting myeloma cells synthesizing only L chains (29). In these hybrids, μ chains were expressed on the cell surface, presumably as result of assembly with the heterologous L chains. However, one must also consider the possibility that in addition to L chains, the myeloma cells may provide other factors necessary for the proper processing of μ chains. Although experiments in which pre-B cells are transformed with L chain DNA have not yet been reported, experiments conducted with IgG suggest that the provision of L chain alone is sufficient for μ chain expression (30). In these experiments, *Xenopus* oocytes were microinjected with mRNA for μ chain, L chain, or μ and L chains together. When μ chain or L chain mRNA is provided alone, the polypeptides accumulate in the cell, but neither is detected in the medium. However, when both mRNAs are injected, covalently assembled monomers are secreted. These results indicate that unless μ and L chains are assembled, they do not gain access to the secretory pathway.

Other investigators have reported that in some transformed cell lines, small amounts of μ chain may become externalized in the absence of L chains (31–33). However, these forms of μ chain appear to be underglycosylated, suggesting that they are altered in some way or that their pathways to the cell surface are different from those followed by covalently assembled IgM. The mechanisms by which assembly may alter the fate of μ chains are as yet unknown. One mechanism may involve the protection of the μ chain from intracellular degradation. Support for this possibility comes from findings that: 1) μ chains assembled with L chains are more resistant to *in vitro* trypsinization than their unassociated counterparts (25); and 2) the turnover of unassembled μ *in vivo* is faster than that of fully assembled μ (31, 34).

In resting B cells, although both μm and μs chains are synthesized, only mIgM is expressed and no μs is secreted. The defect in μs secretion is not in assembly with L chains, since $(\mu sL)_2$ monomer is formed (25). In addition, the monomers are short-lived and the μ chains do not undergo terminal processing. Therefore, for μs, association with an L chain is insufficient to guarantee either intracellular transport or stabilization against catabolism. Sidman (13) has identified an intermediate in μs processing that is present in secreting cells, but not in resting B cells. The intermediate is smaller than other newly synthesized or terminally processed μs due to differences in the oligosaccharide chains. These findings indicate that μs fails to undergo certain trimming reactions in nonsecretory cells, and they are consistent with a block in transit of μs to the Golgi.

Circumstantial evidence suggests that polymerization of monomer and J chains is important in control of μs secretion. Acquisition of the ability to secrete IgM in plasma cells is closely correlated with the onset of J chain production (35). Furthermore, fusion of cells of the mIgM phenotype with cells that produce J chains, but not IgM, results in heterokaryons that secrete pentameric sIgM (36). The J chain may therefore be necessary for secretion, but the possibility that other factors supplied by the secreting parent are responsible for secretion in the hybrid must also be considered. This question could be resolved in experiments in which resting B cells are induced to synthesize J chains by transfection with J chain DNA. Such experiments have not yet been reported.

Class I Histocompatibility Antigens

The class I major histocompatibility antigens HLA-A, B, and C in humans and H-2D, K, and L in the mouse are good models for the study of cell surface oligomeric proteins. They are widely distributed, present on all nucleated cell types, and are particularly abundant on the surface of some lymphoblastoid cell lines (37). A number of specific antisera are available to facilitate rapid purification of the antigens. And finally, the antigens undergo a series of cotranslational and posttranslational modifications that are prototypic for those of a broad class of glycoproteins.

The class I antigens are heterodimers (see Figure 5.1) composed of a highly polymorphic μ chain (molecular weight, 40,000 to 44,000) in noncovalent association with an L chain, β_2 microglobulin (β_2M) (molecular weight, 12,000) (38). The μ chain spans the plasma membrane, and 80% of the chain, including the amino terminus, is extracellular (39). The β_2M chain, on the other hand, is completely extracellular and is bound to the cell surface only by its association with the integral μ chain (40, 41).

The μ and L chains are encoded by separate genes and are synthesized from separate mRNAs (42). Both polypeptides are synthesized on membrane-bound polysomes and are cotranslationally translocated into the cisternae of the ER (42). Each chain is synthesized with an amino terminal signal sequence that is cotranslationally cleaved (43). In the case of β_2M, the entire chain is translocated into the cisternae. For the μ chain, though, synthesis of a hydrophobic stretch of 25 to 30 amino acids halts translocation, and the C terminal domain remains cytoplasmic (44, 45). Another distinction between the μ and β_2M chains is that only the μ chains are glycosylated. From one to three high-mannose core oligosaccharides are cotranslationally linked to asparagine residues (46, 47). As discussed below, these oligosaccharides are subsequently processed to the complex type (26, 43).

Assembly of the dimer proceeds rapidly after synthesis of the μ chain. In pulse chase experiments with cells of lymphoid origin, most of the newly synthesized μ chain is assembled within 5 to 10 minutes (45, 48). In contrast, $\beta_2 M$ is found in excess over μ chains. Free $\beta_2 M$ persists in labeled cells and can be secreted into the medium as a free monomer (42, 49).

Dramatic changes in the conformation of the μ chain accompany heterodimer formation. Populations of unassembled and assembled heavy chains can be clearly distinguished by antibodies generated against either denatured or native antigen (48).

Assembly and the accompanying conformational changes do not require information encoded in the transmembrane or cytoplasmic domain. In gene transfer studies, the cytoplasmic domain of an H-2 μ chain was deleted *in vitro* and the mutant gene was expressed in mouse L cells. The altered μ chains assembled with $\beta_2 M$ and acquired the antigenic properties of the normal surfce antigen (50).

Using another approach, mutant human cell lines were selected for their inability to express HLA-A antigen on the cell surface. A class of these mutants was found to synthesize a μ chain variant that—presumably due to a deletion of the transmembrane domain—was not anchored in the plasma membrane. The unanchored μ chain did assemble with $\beta_2 M$ and became conformationally mature. The resulting antigen was secreted into the medium, rather than remaining on the cell surface (51). Together, these results indicate that only the extracellular domains of the μ chain and $\beta_2 M$ are involved in the assembly and transport processes.

One covalent modification that is known to take place in the extracellular domain of the class I antigens is the formation of intrachain disulfide bonds. The role of a disulfide bond in the C_1 domain of the μ chain has been investigated by *in vitro* mutagenesis. Substitution of a serine for cysteine-101 results in loss of most, but not all, allodeterminants, as well as complete loss of recognition by alloreactive killer T cells (52). Since reactivity with some antibodies is retained, it would appear that this mutation disrupts many conformational-dependent epitopes, but does not block completely the cell surface expression of the molecule.

Transit of the class I heterodimer from the ER through the Golgi compartment to the cell surface is coincident with changes in the structure of the oligosaccharide chains (26). By analogy with other glycoproteins, the class I antigen glycans presumably undergo trimming in the ER and Golgi, and then terminal sugars are added in the late Golgi just prior to appearance on the cell surface. Experiments with tunicamycin, an inhibitor of N glycosylation, have demonstrated that core glycosylation is not critical to either assembly or transport of the HLA and H-2 molecules (45, 53). However, changes in the oligosaccharide chains can serve as convenient markers for progress of the protein along the transit and processing pathways. In particular, resistance to endoglycosidase H digestion, as manifested by unaltered mobility on SDS-polyacrylamide gels, is used as an indication that a glycoprotein has reached the late Golgi in the proper conformation for addition of terminal sugars.

Assembly of the oligomer has been shown to have a distinct effect on transport of the antigen to the cell surface. The first evidence for this conclusion developed from the observation that the human cell line Daudi did not present HLA antigens on the cell surface (54). It was subsequently shown that Daudi cells do not synthesize β_2M (55), but do synthesize HLA μ chains (43). The μ chains remain intracellular and do not acquire endoglycosidase H resistance (56). When Daudi cells are fused with cell lines that do synthesize β_2M, HLA antigens corresponding to the Daudi-synthesized μ chains are expressed on the cell surface (57).

Mutant cell lines derived from a human lymphoblastoid cell line T5-1 have been isolated in which association of HLA-A chains with β_2M does not occur (58). The lesion appears to be in the primary structure of the μ chain. Although the altered μ chains undergo core glycosylation, the glycan chains do not mature, and therefore remain endo H-sensitive. Furthermore, the unassembled μ chains do not become conformationally mature as determined by immunologic criteria, nor do they reach the cell surface. The failure to reach the surface is not due to a rapid degradation of the defective unassembled polypeptides, since they appear to be stable within the cell. These findings strongly indicate that assembly of the oligomer and/or conformational maturation is critical to the transport of μ chains through the processing pathway to the cell surface. It appears as though the association of μ chain and β_2M and the accompanying conformational changes reveal information that addresses the molecule for its proper destination.

Class II Antigens

Studies of another group of histocompatibility antigens, the class II antigens, have revealed some novel biosynthetic features with regard to the assembly process. The class II antigens include, among others, the HLA-DR antigens in humans and the Ia antigens in the mouse (59). On the cell surface, the antigens are noncovalently linked dimers (see Figure 5.1) composed of an α chain (molecular weight, 33,000 to 35,000) and a β chain (molecular weight, 27,000 to 29,000). Both subunits span the plasma membrane and have small carboxy terminal cytoplasmic domains (39).

A unique feature of the biosynthesis of the HLA-DR and Ia antigens involves the transient association of the α- and β-subunits with a third polypeptide called the γ or I chain (molecular weight, 31,000 to 33,000) (59). The I chain is found in association with α and β within the cell, but not on the cell surface. This unusual temporary association has led to the proposal that the I chain may facilitate $\alpha\beta$ assembly and/or transit of the complex to the cell surface, (60, 61). However, it is equally possible that it is association of I with α or β that is required for I chain transport. These possibilities will be discussed below.

In many respects, the synthesis of the α and β chains of the class II antigens resembles that of the μ chains of the class I type. They are synthesized from separate mRNAs that are encoded with the major histocompatibility complex (59, 62). Both chains are synthesized with N terminal signal peptides that are cleaved, and both undergo N-linked core glycosylation (63). The N-linked core oligosaccharides can be processed to the complex type or may remain high-mannose type. In the case of HLA-DR α chains, both types of oligosaccharide are present on one chain (60, 64).

Biosynthesis of the I chain differs in several respects. The I chain is not encoded within the major histocompatibility complex (MHC), and it is not synthesized with a cleavable, N terminal leader sequence (65, 66). Although I chains also span the microsomal membrane, they possess an unusual membrane orientation (67, 68). The N terminus remains on the cytoplasmic side, while the C terminus is translocated to the luminal side. And finally, the I chain has both N- and O-linked oligosaccharide chains (69).

Newly synthesized α- and β-subunits associate rapidly with I chains in the RER, both *in vivo* and in an *in vitro* system supplemented with microsomal membranes (61). The I chain appears to be synthesized in excess of α- and β-subunits, and may be assembled from a large and stable intracellular pool of monomers. Identification of an αI assembly intermediate implies an ordered addition, but it is not known if formation of the αI is necessary for subsequent $\alpha\beta$ formation. Using a number of different antibody specificities, it has been possible to demonstrate that α and β subunits undergo several conformational changes following synthesis, and before cell surface expression (60, 61, 70). However, the exact intracellular compartment in which $\alpha\beta$ assembly occurs has not been determined. As with class I antigens, assembly of the $\alpha\beta$ complex appears to be necessary for transport of polypeptide chains to the cell surface, because in mouse strains that are defective in synthesis of the α-subunit, the β chain remains intracellular (71, 72).

All three subunits undergo transit through the Golgi apparatus. Some N-linked oligosaccharides of the α- and β-subunits undergo maturation to the complex type (60). In addition, the O-linked oligosaccharides are added to the I chain (69). Under normal conditions, the O-linked chains become highly modified with sialic acid. At some point between the addition of the terminal sialic acids and appearance of the $\alpha\beta$ complex on the cell surface, the I chain dissociates. The fate of the dissociated I chain is unknown, but some of it remains detectable on the cell surface, probably in the form of disulfide-linked dimers (73). No function has been attributed to the I chain.

Monensin, a carboxylic ionophore known to disrupt intracellular traffic, produces interesting effects on HLA-DR biosynthesis. In treated cells, the α- and β-subunits appear to become mature in a normal fashion; their oligosaccharides become endo H-resistant (69). However, the I chains do not become properly modified with sialic acid, as demonstrated by their unusual mobility in 2D gel electrophoresis and the lack of effect of neuraminidase on their mobility. The $\alpha\beta$I complex accumulates intracellularly. Apparently, monensin blocks transport of the complex before the addition of the final sugars to the O-linked, but not to the N-linked, oligosaccharides of α and β. Another effect of monensin is that the dissociation of I from the $\alpha\beta$ complex is blocked. This failure to dissociate may be due to the failure of the O-linked oligosaccharides to mature or to some other process that occurs distal to the monensin block.

Recently, gene transfer experiments using cloned genes for α- and β-subunits expressed in mouse L cells have shown that functional HLA class II antigens can be expressed in the absence of expression of exogenous I chain genes (74, 75). If no endogenous I chain gene is expressed under these conditions, it would indicate that the I chain is not required for α- and β-subunit transport or assembly, thus again raising the possibility that αI or βI assembly is necessary for I chain transport.

Acetylcholine Receptor

Although the nicotinic acetylcholine receptor (AChR) of the neuromuscular junction is perhaps the best-described example of a transmembrane ion channel, its structure and biosynthesis are less well understood than those of the previous models. At least part of this difference is accounted for by the added level of structural complexity. The receptor is a pentamer (molecular weight, 250,000) composed of four homologous, but nonidentical, integral membrane subunits in a stoichiometry of $\alpha_2\beta\gamma\delta$ (Figure 5.2). The complete primary structure of all four subunits of the electric ray torpedo has been determined from the cDNA sequences (76). In addition, complete sequence information is available for three of the four bovine subunits (77–79) and human α-subunit (77). Predictions of secondary structure suggest that each subunit may span the membrane bilayer four or five times (80, 81), while the N termini have been localized to the extracellular side (82) and the C termini to the cytoplasmic side of the plasma membrane (83, 106). The α- and β-subunits each have a single "high-mannose" N-linked oligosaccharide (84), while γ and δ may have as many as three chains, of which at least one must be of the "complex" type (85, 86).

Two further posttranslational modifications of the ACh receptor have been characterized. The α-subunit has at least one intrachain disulfide bond that has

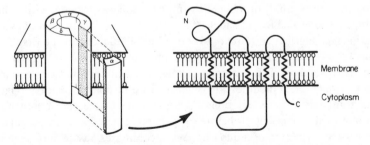

Fig. 5.2. Structure of the native oligomeric ACh receptor, and the extended polypeptide of one of the subunits showing one model of its transmembrane orientation. (See [107].)

been shown to be in close proximity to the ACh-binding site (87). Finally, the ACh receptor subunits acquire covalently linked fatty acid (88).

A cell culture system for the study of synthesis and assembly of torpedo ACh receptor is not available—torpedo cell culture being extremely uncommon. Several mammalian and avian skeletal muscle culture systems are useful for such studies. However, a less complete set of monoclonal antibody reagents are available for the receptors of these higher vertebrates.

Many features of AChR biosynthesis are shared with those of IgM, HLA, H-2, and other glycoproteins. Each of the four subunits is synthesized from a separate mRNA on membrane-bound polysomes (89, 90). Membrane attachment of nascent polysomes is mediated by the signal recognition particle (SRP) (82). Each subunit is synthesized with an N terminal signal sequence that is cotranslationally cleaved, and each undergoes cotranslational glycosylation. Finally, the newly synthesized receptors transit the Golgi apparatus en route to the cell surface (91).

The unique aspects of AChR biosynthesis have been revealed in studies of embryonic muscle cultures. In both chick embryonic muscle (92) and the BC3H-1 cell line (93), new AChR receptors exhibit an unusually long transit time to the cell surface, 2 to 3 hours. This may reflect the complexity of posttranslational modifications or the relative inefficiency of intramembrane transport in muscle. Slow transit times have also been reported for fibronectin secretion from muscle cells (94). As in the synthesis of HLA-DR, ACh receptor subunits undergo a number of conformational changes before the mature structure is achieved. After cotranslational glycosylation and signal peptide cleavage, the newly synthesized subunits undergo a conformational change, which for an α-subunit has been defined as the ability to bind the snake venom α-bungarotoxin. This event in conformational maturation involves no change in sedimentation coefficient and occurs before $\alpha\beta$-subunit association (96). However, because this conformational change is not observed in a cell-free protein-synthesizing system supplemented or unsupplemented with microsomal membranes, it may require a covalent modification not normally performed by such a cell-free system (96).

A second conformational changes involves the assembly of subunits to form the mature oligomeric complex. Oligomeric assembly is detected by the coimmunoprecipitation of heterologous subunits with subunit-specific monoclonal antibodies. The assembly of subunits by this criterion is coincident with a change in sedimentation value from 5S to 9S (95). Although the kinetics of the two conformational changes are different (t-1/2 of 15 to 30 minutes for the first conformational maturation of individual subunits, and t-1/2 of 60 minutes for oligomeric assembly), it is not known yet whether either change is coincident with a known covalent modification, such as disulfide bond formation, oligosaccharide trimming, or fatty acylation. It is not known whether one or both of these conformational changes occur before the subunits reach the Golgi. Finally, evidence for the formation of homo-oligomeric complexes has been obtained in a cell-free system (98), but their relevance to the normal process of assembly is not known (98).

Several examples of stimuli that regulate ACh receptor expression by muscle cells in culture have been described. In BC3H-1 cells, treatment with tunicamycin results in the failure of a newly synthesized α-subunit to acquire α-bungarotoxin-binding activity—an indication that the first conformational change does not occur (84). Although this suggests that glycosylation is necessary for the conformational maturation, it clearly can not be sufficient. The α-subunit synthesized in a cell-free membrane-supplemented system, in which core glycosylation does occur, also does not acquire toxin-binding activity (89, 96).

When BC3H-1 cells are kept in log phase growth, they have very few ACh receptors on their surface, whereas fully differentiated cells accumulate approximately 10^6 receptors per cell (93). Olson *et al.* showed that differentiation can be induced by serum step-down and reversed by readdition of serum (99). Thus, it is possible to both induce ACh receptor expression and deinduce sequentially in the same cultures simply by changes of culture medium. Surprisingly, it was found that ACh receptor α-subunit synthesis does not correlate with the synthesis of new receptor oligomers. Uninduced, induced, and deinduced BC3H-1 cells synthesize significant levels of α-subunit, whereas the incorporation of the newly synthesized α-subunit into a 9S oligomeric complex occurred only in induced cultures (99). Furthermore, even in differentiated cultures, the fraction of newly synthesized α that remains unassembled is high, 40 to 70% (100, 101). This behavior is consistent with regulation of a posttranslational step involved in subunit conformational maturation or assembly; or alternatively, with the stringent regulation of synthesis of γ or δ receptor subunits that can not be currently measured. The resolution of these two alternatives must await the development of additional monoclonal antibody reagents.

The fate of unassembled subunits in BC3H-1 cells is rapid degradation. In fact, degradation is so rapid that subunit synthesis is difficult to detect in undifferentiated cultures in which no subunits are stabilized by assembly. In these instances, it has been necessary to use short pulse-labeling experiments to detect α- and β-subunit synthesis (99, 101). Similar effects were noted when torpedo α-subunit was expressed in the absence of other subunits in xenopus oocytes that were microinjected with α-subunit mRNA prepared from cells transfected with a recombinant expression vector (102).

Primary cultures of rat embryo muscle cells offer the possibility of studying regulation of ACh receptor synthesis mediated by contractile activity. These cells have been shown to initiate spontaneously sustained, high-frequency action potentials, resulting in contraction. Under these conditions, several enzymes of energy metabolism and ACh receptors are down-regulated. Specific inhibition of these spontaneous action potentials with the Na^+ channel blocker tetrodotoxin results in up-regulation (103). In this system, the activity-linked inhibition of new receptor synthesis has been found to be both at the level of subunit synthesis and at the level of subunit assembly (105, 106). As in BC3H-1 cells, it can not be determined yet whether the apparent block in α-subunit assembly is due to regulation of the assembly process *per se*, or whether γ- and/or δ-subunit synthesis is stringently regulated. However, unlike in BC3H-1 cells, the unassembled α-subunit in primary rat embryo muscle cells is metabolically stable; and surprisingly, it acquires high-affinity α-bungarotoxin-binding activity. Furthermore, since the sedimentation coefficient of this species is 5S, it has been possible to show that it is not transported to the cell surface (105, 106). Thus, primary cultures of rat muscle cells offer the possibility of studying the mechanism of activity-linked regulation of ACh receptor assembly, as well as the properties of the unassembled α-subunit.

Conclusions

Some similarities, and thus some answers to the question posed in the introductory paragraph, can be arrived at from a composite look at the preceding four examples. Obviously, the number of well-characterized multisubunit membrane proteins is too few, and the diversity of structure among them is too great, to suggest specific mechanisms involved in assembly. However, the comparison seems useful, because it indicates the directions in which future experiments may provide the most relevant information.

Assembly does seem to be an inefficient process. For each of the proteins considered, one or more subunits is produced in excess of the amount finally incorporated into the mature oligomer. In many cases, the unassembled species are rapidly degraded, resulting in a low steady-state level of free subunit. In fact, rapid and selective degradation of free subunit may make it difficult to determine accurately the efficiency of assembly.

No clear and direct evidence for the existence of an active process of assembly has been found. However, this possibility is difficult to exclude unless all forms of assembly intermediates of all subunits can be accounted for in a quantitative manner. Assembly, in some cases, may be dependent upon a conformational change that, in turn, may be dependent upon a covalent modification. Covalent modifications can be relatively inefficient, depending upon the cell and tissue types in question; and if required even indirectly for assembly, they may serve to regulate the synthesis of biologically active oligomers.

The site of assembly appears to be the RER, although assembly in the Golgi can not be ruled out. If, for some proteins, assembly does occur in the Golgi, that process may require Golgi-specific modifications. In the case of proteins synthesized in low abundance, the Golgi may play a passive role by generating higher local concentrations of subunits. Whether in the Golgi or the RER, assembly involves significant changes in the conformation of constituent subunits; and the final conformation appears necessary for correct, efficient cell surface expression. Although far from proven, it would appear that correct intracellular transport of integral membrane proteins is dependent upon the native conformation of the large hydrophilic, extracellular domains. At present, it is difficult to conceive of how the native conformations of the extracellular domains of proteins as diverse as IgM and AChR could influence their transport to correct subcellular location. Resolution of this issue should prove to be an interesting challenge.

Acknowledgments. Work from the authors' laboratory was supported by grants from the NIH and The Muscular Dystrophy Associations of America Inc. Barry Carlin was a postdoctoral fellow of The Muscular Dystrophy Associations of America Inc.

References

1. Walter, P., Gilmore, R. & Blobel, G. (1984) Cell *38*, 5–8.
2. Rothman, J.E. & Lenard, J. (1984) TIBS *9*, 176–178.
3. Wall, R. & Kuehl, M. (1983) Ann. Rev. Immunol. *1*, 393–422.
4. Potter, M. (1972) Physiol. Rev. *52*, 632–719.
5. Cathou, R.E. (1978) In Immunoglobulins, 37–84, Litman, G.W. & Good, R.A., eds. Plenum Press, New York.
6. Roger, J. & Wall, R. (1984) Adv. Immunol. *35*, 39–59.
7. Metzger, H. (1970) Adv. Immunol. *12*, 57–116.
8. Koshland, M.E. (1975) Adv. Immunol. *20*, 41–69.
9. Vassalli, P., Tartakoff, A., Pink, J.R.L. & Jatton, J.C. (1980) J. Biol. Chem. *255*, 11822–11827.
10. Singer, P.A., Singer, H.H. & Williamson, A.R. (1980) Nature *285*, 294–300.
11. Rogers, J., Early, P., Carter, C., Calame, K., Bond, M., Hood, L. & Wall, R. (1980) Cell *20*, 303–312.
12. Alt, F.B., Bothwell, A.L.M., Knapp, M., Siden, E., Mather, E., Koshland, M. & Baltimore, D. (1980) Cell *20*, 293–301.
13. Sidman, C. (1981) Cell *23*, 379–389.
14. Kehry, M., Sibley, C., Fuhrman, J., Schilling, J. & Hood, L.E. (1980) Cell *21*, 393–406.
15. Bergman, L.W. & Kuehl, W.M. (1982) In The Glyco Conjugates, 82–98, Horowitz, M.I., ed. Academic Press, New York.
16. Tartakoff, A., Hoessli, D. & Vassalli, P. (1981) J. Mol. Biol. *150*, 525–535.
17. Kuehl, W.M. (1977) Curr. Topics Microbiol. Immunol. *76*, 1–47.
18. Bergman, L.W. & Kuehl, W.M. (1978) Biochemistry *17*, 5174–5180.
19. Tartakoff, A. & Vassalli, P. (1979) J. Cell Biol. *83*, 284–299.
20. Dulis, B.H. (1983) J. Biol. Chem. *258*, 2181–2187.
21. Sidman, C., Potash, M.J. & Kohler, G. (1981) J. Biol. Chem. *256*, 13180–13187.
22. Kubo, R. & Pelanne, M.L. (1983) Mol. Immunol. *20*, 67–76.
23. Sibley, C.H. & Wagher, R.A. (1981) J. Immunol. *126*, 1868–1873.
24. Tartakoff, A.M. (1983) Cell *32*, 1026–1028.
25. Tartakoff, A.M. & Vassalli, P. (1977) J. Exp. Med. *146*, 1332–1349.
26. Tartakoff, A., Hoessli, D. & Vassalli, P. (1981) J. Mol. Biol. *150*, 525–535.
27. Siden, E., Alt, F.W., Sato, V. & Baltimore, D. (1981) PNAS *78*, 1823–1827.
28. Mains, P.E. & Sibley, C.H. (1983) J. Biol. Chem. *258*, 5027–5033.
29. McCune, J.M. & Fu, S.M. (1981) J. Immunol. *127*, 2609–2611.
30. Valle, G., Besley, J. & Colman, A. (1981) Nature *291*, 338–340.
31. Hendershot, L. & Levitt, D. (1984) J. Immunol. *132*, 502–509.
32. Paige, C.J., Kincarde, P.W. & Ralph, P. (1981) Nature *292*, 631–633.
33. Levitt, D. & Cooper, M. (1980) Cell *19*, 617–625.

34. Dulis, B.H., Kloppel, T.M., Grey, H.M. & Kubo, R.T. (1982) J. Biol. Chem. *257*, 4369-4374.
35. McHugh, Y., Yagi, M. & Koshland, M.E. (1981) In B Lymphocytes in the Immune Response, 467-474, Klinman, N., Mosier, D., Scher, I. & Vitetta, E., eds. Elsevier, Holland.
36. Raschke, W.C., Mather, E.L. & Koshland, M.E. (1979) PNAS *76*, 3469-3473.
37. Klein, J., Figueroa, F. & Nagy, Z.A. (1983) Ann. Rev. Immunol. *1*, 119-142.
38. Dobberstein, B., Kvist, S. & Roberts, L. (1982) Phil. Trans R. Soc. Land. *300*, 161-172.
39. Walsh, F.S. & Crumpton, M.J. (1977) Nature *269*, 307-311.
40. Peterson, P.A., Rask, L. & Lindblom, J.B. (1974) PNAS *71*, 35-39.
41. Rask, L., Lindblom, J.B. & Peterson, P.A. (1974) Nature *249*, 833-834.
42. Dobberstein, B., Garoff, H., Warren, G. & Robinson, P.S. (1979) Cell *17*, 759-769.
43. Ploegh, H.L., Cannon, L.E. & Strominger, J.L. (1979) PNAS *76*, 2273-2277.
44. Nathenson, S.G., Vehara, H., Ewenstein, B.M., Kindt, T.J. & Coligan, J.E. (1981) Ann. Rev. Biochem. *50*, 1025-1051.
45. Owen, M.J., Kissonerghis, A. & Lodish, H.F. (1980) J. Biol. Chem. *255*, 9678-9684.
46. Parham, P., Alpert, B.N., Orr, H.T. & Strominger, J.L. (1977) J. Biol. Chem. *252*, 7555-7567.
47. Coligan, J.E., Kindt, T.J., Vehara, H., Martinko, J. & Nathenson, S.G. (1981) Nature *291*, 35-39.
48. Krangel, M.S., Orr, H.T. & Strominger, J.L. (1979) Cell *18*, 979-991.
49. Algranati, I.D., Milstein, C. & Ziegler, A. (1980) Eur. J. Biochem. *103*, 197-207.
50. Zuniga, M., Malissen, B., McMillan, M., Brayton, P.R., Clark, S.S., Forman, J.F. & Hood, L. (1983) *34*, 535-544.
51. Krangel, M.S., Pious, D. & Strominger, L.L. (1984) J. Immunol. *132*, 2984-2991.
52. Shiroishi, T., Evans, G.A., Appella, E. & Ozato, K. (1984) PNAS *81*, 7544-7548.
53. Ploegh, H., Orr, H.T. & Strominger, J.L. (1981) J. Immunol. *126*, 270-275.
54. Evrin, P.E. & Nilsson, K. (1974) J. Immunol. *112*, 137-144.
55. Nilsson, K., Ervin, P. & Welsh, K.I. (1974) Transplant. Rev. *21*, 53-84.
56. Sege, K., Rask, L. & Peterson, P.A. (1981) Biochemistry *20*, 4523-4530.
57. Klein, G., Terasaki, P., Billing, R., Jondal, M., Rosen, A., Zeuthen, J. & Clements, G. (1977) Int. J. Cancer *19*, 66-76.
58. Krangel, M.S., Pious, D. & Strominger, J.L. (1982) J. Biol. Chem. *257*, 5296-5305.
59. Kaufman, J.F., Auffray, C., Korman, A.J., Shackelford, D.A. & Strominger, J. (1984) Cell *36*, 1-13.
60. Owen, M.J., Kissaneughis, A.M., Lodish, H.F. & Crumpton, M.J. (1981) J. Biol. Chem. *256*, 8987-8993.
61. Kvist, S., Wiman, K., Claesson, L., Peterson, P.A. & Dobberstein, B. (1982) Cell *29*, 61-69.
62. Lee, J.S., Trowsdale, J., Travers, P.J., Carey, J., Grosveld, F., Jenkins, J. & Bodmer, W.F. (1982) Nature *299*, 750-752.
63. Korman, A.J., Ploegh, H.L., Kaufman, J.F., Owen, M.J. & Strominger, J.L. (1980) J. Exp. Med. *152*, 65-82.
64. Shackelford, D.A. & Strominger, J.L. (1983) J. Immunol. *130*, 274-282.
65. Day, C.E. & Jones, P.P. (1983) Nature *302*, 157-159.
66. Long, E.O., Strubin, M., Wake, C.T., Gross, N., Carrel, S., Goodfellow, P., Accolla, R.S. & Mach, B. (1983) PNAS *80*, 5714-5718.

67. Strubin, M., Mach, B. & Long, E.O. (1984) EMBO J. *3*, 869–972.
68. Singer, P.A., Laver, W., Dembic, Z., Mayer, W.E., Lipp, J., Koch, N., Hammerling, G., Klein, J. & Dobberstein, B. (1984) EMBO J. *3*, 873–877.
69. Machmer, C.E. & Creswell, P. (1982) J. Immunol. *129*, 2564–2569.
70. Shackelford, D.A., Lampson, L.A. & Strominger, J.L. (1981) J. Immunol. *127*, 1403–1410.
71. Jones, P.P. (1980) J. Exp. Med. *152*, 1453–1458.
72. Mathis, D.J., Benoist, C., Williams, V.E., Kanter, M. & McDevitt, H.O. (1983) PNAS *80*, 273–277.
73. Malissen, B., Steinmetz, M., McMillan, M., Pierres, M. & Hood, L. (1983) Nature *305*, 440–443.
74. Malissen, B., Peele-Price, M., Govermann, J.M., McMillan, M., White, J., Kappler, J., Marrack, P., Pierres, A., Pierres, M. & Hood, L. (1984) Cell *36*, 319–327.
75. Rabourdin-Combe, C. & Mach, B. (1983) Nature *303*, 670–674.
76. Numa, S., Noda, M., Takahashi, H., Tanabe, T., Toyosato, M., Furutani, Y. & Kikyotani, S. (1983) Cold Spring Harbor Symp. Quant. Biol. *48*, 57–69.
77. Noda, M., Furutani, Y., Takahashi, H., Toyosato, M., Tanabe, T., Shimizu, S., Kikyotani, S., Kayano, T., Hirose, T., Inayama, S. & Numa, S. (1983) Nature *305*, 818–823.
78. Takai, T., Noda, M., Furutani, Y., Takahashi, H., Notaki, M., Shimizer, S., Kayano, T., Tanabe, T., Tanaka, K., Hirose, T., Inayama, S. & Numa, S. (1984) Eur. J. Biochem. *143*, 109–115.
79. Tanabe, T., Noda, M., Furutani, Y., Takai, T., Takahashi, H., Tanaka, K., Hirose, T., Inayama, S. & Numa, S. (1984) Eur. J. Biochem. *144*, 11–17.
80. Finer-Moore, J. & Stroud, R.M. (1984) Proc. Natl. Acad. Sci. USA *81*, 155–159.
81. Guy, H.R. (1984) Biophys. J. *45*, 249–261.
82. Anderson, D.J., Walter, P. & Blobel, G. (1982) J. Cell Biol. *93*, 501–506.
83. Ratnam, M. & Lindstrom, J. (1984) Biochem. Biophys. Res. Comm. *122*, 1225–1233.
84. Merlie, J.P., Sebbane, R., Tzartos, S. & Lindstrom, J. (1982) J. Biol. Chem. *257*, 2694–2701.
85. Lindstrom, J., Merlie, J. & Yogeeswaran, G. (1979) Biochemistry *18*, 4465–4470.
86. Meunier, J.C., Sealock, R., Olsen, R. & Changeux, J.P. (1974) Eur. J. Biochem. *45*, 371–394.
87. Kao, P.N., Dwork, A.J., Kaldany, R.J., Silver, M.L., Wideman, J., Stein, S. & Karlin, A. (1984) J. Biol. Chem. *259*, 11662–11665.
88. Olson, E.N., Glaser, L. & Merlie, J.P. (1984) J. Biol. Chem. *259*, 5364–5367.
89. Anderson, D.J. & Blobel, G. (1981) Proc. Natl. Acad. Sci. USA *78*, 5598–5602.
90. Merlie, J.P., Hofler, J.G. & Sebbane, R. (1981) J. Biol. Chem. *256*, 6995–6999.
91. Fambrough, D.M. & Devreotes, P.N. (1978) J. Cell Biol. *76*, 237–244.
92. Devreotes, P.N. & Fambrough, D.M. (1975) J. Cell Biol. *65*, 335–358.
93. Patrick, J., McMillan, J., Wolfson, H. & O'Brien, J.C. (1977) J. Biol. Chem. *252*, 2143–2153.
94. Gardner, J.M. & Fambrough, D.M. (1983) J. Cell Biol. *96*, 474–485.
95. Merlie, J.P. & Lindstrom, J. (1983) Cell *34*, 747–757.
96. Sebbane, R., Clokey, G., Merlie, J.P., Tzartos, S. & Lindstrom, J. (1983) J. Biol. Chem. *258*, 3294–3303.
97. Anderson, D.J. & Blobel, G. (1983) Proc. Natl. Acad. Sci. USA *80*, 4359–4363.
98. Merlie, J.P. (1984) Cell *36*, 573–575.

99. Olson, E.N., Glaser, L., Merlie, J.P., Sebbane, R. & Lindstrom, J. (1983) J. Biol. Chem. *258*, 13946–13953.
100. Merlie, J.P., Sebbane, R., Gardner, S., Olson, E. & Lindstrom, J. (1983) Cold Spring Harbor Symp. Quant. Biol. *48*, 135–145.
101. Olson, E.N., Glaser, L., Merlie, J.P. & Lindstrom, J. (1984) J. Biol. Chem. *259*, 3330–3336.
102. Mishina, M., Kurosaki, T., Tobimatsu, T., Morimoto, Y., Noda, M., Yamamoto, Y., Terao, M., Lindstrom, J., Takahashi, T., Kuno, M. & Numa, S. (1984) Nature *307*, 604–608.
103. Lawrence, J.C. & Salsgiver, W.J. (1983) Am. J. Physiol. *244*, C348–355.
104. Carlin, B.E., Lawrence, J.C., Jr., Lindstrom, J.M. & Merlie, J.P. (1986) Proc. Natl. Acad. Sci. USA, *83*, 498–502.
105. Carlin, B.E., Lawrence, J.C., Jr., Lindstrom, J.M. & Merlie, J.P. (1986) J. Biol. Chem. in press.
106. Young, E.F., Ralston, E., Blake, J., Ramachandran, J., Hall, Z.W. & Stroud, R.M. (1985) Proc. Natl. Acad. Sci. USA *82*, 626–630.

6

Structure, Function, and Biosynthesis of Fatty Acid-Acylated Proteins

Eric N. Olson

The covalent attachment of long chain fatty acids to proteins was first described for brain myelin proteolipoprotein (15). The major membrane lipoprotein in the *Escherichia coli* cell wall was later shown to contain fatty acids attached through both ester and amide linkages (6, 19). This covalent modification has been subsequently shown to be common to a variety of eukaryotic, bacterial, and viral membrane glycoproteins. However, fatty acid acylation is a highly selective modification, even among membrane glycoproteins, since only a subset of membrane-associated proteins contain lipids. Because fatty acylation is such a recently identified covalent modification, the characteristics of a protein that specify whether it becomes acylated, as well as the biosynthetic events involved in the attachment of fatty acids to proteins, are only beginning to be explored. This review will summarize current knowledge regarding this unique covalent modification and will discuss possible avenues of future research that might provide a clearer understanding of the structure, function, and biosynthesis of fatty acid-acylated proteins.

Proteins containing covalently bound fatty acid are generally referred to as "acylproteins" (51). In order for a protein to be classified as an acylprotein, the fatty acid moiety must be resistant to removal with organic solvents, resistant to denaturing conditions, and resistant to removal with proteases. The covalent fatty acid must also be identified following its release from the polypeptide by chemical cleavage of the protein-lipid linkage. A variety of lipid moieties other than fatty acids (*e.g.*, phospholipid, diglyceride, mevalonate) has also been shown to be covalently associated with cellular proteins. However, this review will focus only on proteins that contain covalent fatty acid.

Nature of the Protein-Lipid Linkage

The majority of the acylproteins described to date have been shown to contain fatty acid linked to the polypeptide by an O ester or thiol ester bond. Proteins containing esterified fatty acids are listed in Table 6.1. All of these proteins, with the exception of myelin proteolipoprotein, contain covalent fatty acid linked through a bond that exhibits the characteristics of a thiol ester. Thiol ester linkages are extremely labile and are rapidly broken by treatment with hydroxylamine at neutral pH. The linkage of the fatty acid moiety to myelin proteolipoprotein is considerably more stable than a typical thiol ester, but it can be broken by hydroxylamine treatment at alkaline pH. As will be discussed below, this protein-lipid linkage has been identified as an O ester.

Recently, a small number of proteins has also been reported to contain covalent fatty acid attached through a linkage that exhibits the characteristics of an amide bond. The fatty acid moiety on these acylproteins is highly resistant to removal with hydroxylamine and requires acid hydrolysis for release from the polypeptide. Many of these proteins are soluble, nonglycosylated proteins that transiently associate with membranes (Table 6.2). In those cases in which the amide-linked fatty acid has been identified, it has been shown to be the 14-carbon fatty acid, myristate.

Despite the identification of this wide range of acylproteins, the actual amino acid residues that serve as acylation sites have been determined in only a few cases. The difficulties in identifying acylated amino acids are due, at least in part, to the extremely hydrophobic nature of acylated protein domains, in addition to

Table 6.1. Proteins Containing Fatty Acid Linked by an Ester-Type Bond

Protein	References
Semliki Forest virus glycoproteins	52
Sindbis virus glycoproteins	51
Vesicular stomatitis virus glycoprotein G	50
Influenza virus hemagglutinin HA_2	52
Fowl-plague virus hemagglutinin H_2	52
Newcastle disease virus fusion glycoprotein Fl	52
Corona virus glycoprotein E2	32, 52
LaCross virus glycoprotein G1 and G2	34
Simian virus 40 large T antigen	21
Brain myelin proteolipoprotein	2, 15
The transferrin receptor	38
Milk fat globule membrane butyrophilin and xanthine oxidase	24
Bacteriorhodopsin	33
The major histocompability complex antigens	23
Ca^{2+} ATPase	28
p21-Transforming proteins	54
Human gastric mucus glycoprotein	55, 56
Unidentified cellular proteins	5, 31, 34, 46

Table 6.2. Proteins Containing Fatty Acid Linked by an Amide Bond

Protein	References
p60[src]	12, 53
p15	20
p120	54
Catalytic subunit of cAMP-dependent protein kinase	9
Calcineurin b	1
Nicotinic acetylcholine receptor	35
Escherichia coli lipoprotein	19
NADH cytochrome b$_5$ reductase	39
Unidentified cellular proteins	34, 36

the low level of expression of most cellular acylproteins. For the latter reason, the initial discovery by Schlesinger and coworkers (48, 49) that Sindbis virus E1 and E2 and vesicular stomatitis virus G glycoproteins contain covalent fatty acid led to major advances in the understanding of the structure and biosynthesis of fatty-acylated proteins. The envelope virus glycoproteins are amenable for the study of fatty acylation, because following virus infection, host cell protein synthesis is inhibited and viral proteins are synthesized at levels up to 10^3-fold higher than normal cellular proteins (30). The study of fatty acylation of envelope virus glycoproteins has now been extended to a wide range of viruses, and it appears to be a modification common to at least one membrane glycoprotein of every enveloped RNA virus (48, 51).

Amino acid acylation sites within the envelope virus glycoproteins have been localized to hydrophobic membrane domains of these polypeptides. By protease digestion of membranes from cells infected with Sindbis and vesicular stomatis virus, the fatty acid on E1, E2, and G glycoproteins has been shown to be contained within a carboxy terminal domain (29, 49, 44). In the case of G glycoprotein, this domain contains 64 amino acids and includes the cytoplasmic domain, the transmembrane domain of 20 amino acids, and 14 amino acids on the extracellular side of the membrane-spanning domain. By constructing mutagenized cDNA clones for G protein and studying their expression in eukaryotic cells, Rose *et al.* (45) demonstrated that palmitate was linked to an amino acid within the first 14 residues on the carboxy terminal side of the membrane. This sequence contained a cysteine residue that, when changed to a serine by oligonucleotide-directed mutagenesis, resulted in the expression of nonacylated G protein. Thus, this cysteine residue appears to be the acylation site. However, a direct demonstration that this residue is acylated in the normal protein has not yet been possible.

Identification of acylation sites within cellular acylproteins has been considerably more difficult. Some progress has been made, however, toward localizing acylated amino acids to specific regions of acylproteins. For example, [^{3}H]palmitate is incorporated into the HLA-B and HLA-DR heavy chains of the human B lymphoblastoid cells Jy and T51 (23). Protease digestions localize the acylation sites in these proteins to transmembrane hydrophobic domains, each of which contain a single cysteine residue and no serine or threonine residues. Because palmitate appears to be attached to these proteins via a thiol ester linkage, these cysteines appear to be the acylation sites. Similar protease digestion experiments have localized the acylation site on the transferrin receptor to a region of the molecule that is closely associated with the plasma membrane (38). To date, the only cellular protein, containing esterified fatty acid, in which the acylated amino acid residue has been absolutely identified is brain myelin proteolipoprotein (lipophilin). In this case, fatty acid was shown to be attached to Thr-198, which exists within a hydrophilic segment of the polypeptide (59).

Attempts to identify amino acids that contain amide-linked fatty acid have been considerably more successful, due at least in part to the more hydrophilic nature of this class of proteins. N-acylated amino acid residues have been identified for the catalytic subunit of the cAMP-dependent protein kinase (9), calcineurin b (1), the retrovirus-transforming proteins p15 (20) and p60^{src} (53), and NADH cytochrome b$_5$ reductase (39). In each of these cases, an amino terminal glycine residue was shown to be acylated with myristate. Olson *et al.* (35) reported recently that the α- and β-subunits of the nicotinic acetylcholine (ACh) receptor also contain covalent fatty acid attached through a linkage that exhibits the characteristics of an amide bond. In the case of these polypeptides, however, the acylation sites are probably not located at the amino terminal, since these amino acids have been identified by amino acid sequencing (10); and covalent fatty acid would have prevented their identification. An ϵ amino group on a lysine residue in a domain closely associated with the membrane would appear to be a likely site for N-linked acylation of the receptor subunits. It remains to be determined, however, whether fatty acids are linked to proteins through amide bonds both at the amino terminus and at other regions of the polypeptide chain.

The fact that not all proteins containing glycine at the amino terminus are acylated suggests that there is specificity with respect to proteins that can serve as acceptors for myristate. Currently, there is not enough information available on the identities of myristate-containing proteins to permit identification of common structural features or possible amino acid recognition sequences that might

Table 6.3. Amino Acid Sequences at the Amino Termini of Myristylated Proteins

Protein	Sequence	References
Catalytic subunit of cAMP-dependent protein kinase	Myristyl-Gly-Asn-Ala-Ala-Ala-Ala-Lys	9
Calcineurin b	Myristyl-Gly-Asn-Gln-Ala-Ser-Thr-Pro	1
p15	Myristyl-Gly-Gln-Thr-Val-Thr-Thr-Pro	20
p60^{src}	Myristyl-Gly-Ser-Ser-Lys-Ser-Lys-Pro	53
NADH cytochrome b$_5$ reductase	Myristyl-Gly-Ala-Gln-Leu-Ser-Thr-Leu	39

be required for myristylation. The amino acid sequences adjacent to the myristylated glycine residue in the catalytic subunit of cAMP-dependent protein kinase, calcineurin b, p15, p60^{src}, and NADH cytochrome b$_5$ reductase are shown in Table 6.3. While these sequences do share some similarities in amino acid residues or in types of residues at a given location, there is no obvious concensus sequence common to the four acylproteins; *e.g.*, such as that required for N glycosylation. It will be important in the future to identify the recognition sequence(s) that are required for myristylation and to determine whether myristate is attached to proteins only at amino terminal glycine residues.

Acyl Chain Specificity of Acylation

Until recently, the possibility that fatty acylation might be highly specific with respect to acyl chain length of fatty acids had not been thoroughly examined. Using the mouse muscle cell line, BC3H-1, Olson *et al.* (34) addressed this issue by labeling cultures with [^{3}H]palmitate and [^{3}H]myristate and by examining on NaDodSO$_4$ polyacrylamide gels the spectrum of proteins that were acylated with these fatty acids. As shown in Figure 6.1, palmitate and myristate were incorporated into distinct sets of proteins that appear to be minor protein species, since they do not correspond to major methionine-labeled polypeptides. The most highly labeled palmitate-containing proteins exhibited apparent Mrs ~ 18,000 and 20,000. A number of other cell types have also been shown to contain major [^{3}H]palmitate-labeled proteins of Mr ~ 20,000 (46, 51). The identities of these acylated proteins are unknown.

In BC3H-1 cells, the pattern of proteins that was acylated with myristate was very different from the palmitate-labeling pattern. Myristate was also incorporated into acylproteins to a much greater degree than palmitate. The large number of myristate-containing proteins is surprising, considering that myristate comprises less than 2% of the fatty acids in most cells (25, 26). It is tempting to speculate that the specific cellular function of myristate is as a substrate for acylation of proteins. Recent studies indicate that fatty acids other than palmitate and myristate are also incorporated in a highly specific manner into different cellular proteins (E. Olson, unpublished results).

To determine whether this broad spectrum of acylproteins was common to other cell types, Olson *et al.* (34) also examined the patterns of palmitate- and myristate-labeled proteins in 3T3 mouse fibroblasts in the PC12 cells—a rat pheochromocytoma cell line. As was observed with BC3H-1 cells, palmitate and myristate were incorporated into distinct sets of proteins. The palmitate-labeling pattern was similar in the different cell types. Many of the myristate-containing proteins also appeared to be expressed in all of the cells examined, while others were cell-type-specific, suggesting that they may serve specialized functions.

The dramatic acyl chain specificity observed in the experiments described above suggested that additional specificity might reside within the proteins that serve as substrates for acylation. To further investigate the acyl chain specificity

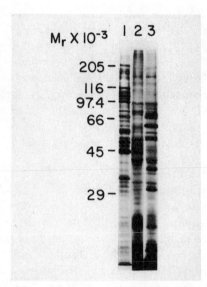

Fig. 6.1. NaDodSO₄ polyacrylamide gel electrophoresis of metabolically labeled proteins. BC3H-1 cells were labeled for 4 h with [³⁵S]methionine (Lane 1), [³H]palmitate (Lane 2), or [³H]myristate (Lane 3). At the end of the labeling period, cell extracts were prepared and analyzed by electrophoresis on a 10% polyacrylamide gel followed by fluorography (34).

of acylation, Olson *et al.* (34) examined the sensitivity of palmitate- and myristate-containing proteins in BC3H-1 cells to alkaline methanolysis and hydroxaminolysis. Treatment of [³H]palmitate-labeled proteins with 1 *M* hydroxylamine, pH 7, resulted in the rapid release of covalent fatty acid, indicating that palmitate was linked to proteins primarily through thiol ester or extremely labile O ester bonds. In contrast to the lability of the palmitate linkage, the bond through which myristate was linked to proteins was resistant to hydroxylamine at pHs 7 and 10, and it required acid hydrolysis for release of fatty acid. These data suggest that myristate is highly specific for amide linkages to proteins. To determine whether palmitate and myristate were the actual fatty acid species attached through ester and amide linkages, respectively, protein-bound lipids were released with methanolic HCl and analyzed by high-pressure liquid chromatography. Analysis of [³H]palmitate-labeled proteins indicated that palmitate was the fatty acid that was incorporated into proteins, and that no interconversion to other fatty acids occurred during the labeling period. In contrast, the fatty acids released from myristate-labeled proteins were primarily myristate, with small quantities of palmitate. The degree of chain elongation of myristate to palmitate varied between experiments, but did not exceed 20% of the protein-bound fatty acid in BC3H-1 cells. A small fraction of the protein-linked myristate in BC3H-1 cells was present in an ester linkage. This fraction varied between different cell types, but in all cells examined, greater than 70% of the myristate was linked to proteins by an amide linkage. The degree of chain elongation of myristate, as well as its conversion into amino acids, also varied significantly between cell types. These studies indicate that metabolic labeling of cellular proteins with fatty acids, followed by NaDodSO$_4$ polyacrylamide gel electrophoresis, is not sufficient for the identification of protein-bound fatty acids—and that chemical characterization of the radioactivity associated with a particular protein is essential. Together, these results also demonstrate that fatty acylation is a very common covalent modification that exhibits a high degree of specificity with respect to fatty acyl donor and acceptor.

Subcellular Distribution of Acylproteins

Until recently, it was generally thought that fatty acid acylation occurred within a membrane subcompartment and was restricted to membrane glycoproteins. The studies of the envelope virus glycoproteins, which led to this conclusion, are discussed in detail in a later section. The demonstration of covalent fatty acid on retrovirus-transforming proteins that are synthesized on free polysomes (27, 54) provided a preliminary indication that acylation might be a modification common to cytoplasmic as well as membrane proteins. To examine the subcellular distribution of cellular acylproteins, Olson et al. (34) labeled BC3H-1 cells separately with [^{3}H]palmitate, and [^{3}H]myristate, and they isolated cytosolic and membrane fractions. Analysis of these fractions by electrophoresis on NaDodSO$_4$ polyacrylamide gels indicated that virtually all of the proteins acylated with palmitate were localized in the membrane fraction. In contrast, myristate-containing proteins were found in both the membrane and cytosolic fractions. Many of the myristate-containing proteins appeared to be distributed exclusively in one or the other of the two fractions, while others were found in both fractions. Similar results for the subcellular distribution of myristylated proteins in chicken embryo fibroblasts were also reported by Buss et al. (8).

The large number of myristate-labeled proteins in the soluble fraction is striking and suggests that fatty acylation is not restricted to proteins that are tightly bound to membranes. It is not known yet whether some of the myristate-containing proteins associated with membranes are synthesized in the cytosol and bind to membranes following acylation, as in the case of the retrovirus transforming proteins (8, 12, 20, 54), or whether these proteins are integral membrane glycoproteins. It is possible that some of the soluble acylproteins may be loosely attached to membranes and are released during the cell fractionation procedure. It also remains to be determined whether the membrane-associated acylproteins are found exclusively in the plasma membrane or, alternatively, are localized to specific intracellular membrane systems or organelles.

The finding that many myristate-containing proteins are localized to the cytosol is surprising, since these proteins would be expected to be extremely hydrophobic and insoluble in an aqueous milieu. The solubility of these cytosolic acylproteins also might be maintained by protein folding, so that the lipid moiety is buried within a cleft or pocket in the native structure of the polypeptide. Alternatively, these soluble acylproteins might be complexed with carrier proteins that maintain their solubility in the cytoplasm. In this regard, studies of $p60^{src}$ have shown that this acylprotein is synthesized on free polysomes, released into the cytosol, and reaches the plasma membrane 5 to 15 minutes thereafter (27). During its brief transit through the cytosol, $p60^{src}$ is associated with two carrier proteins: p89 cell and p50 cell (7, 11). This mechanism for transit of myristate-containing proteins to the plasma membrane may represent a pathway followed by other acylproteins.

Biosynthesis of Acylproteins

The high degree of specificity of palmitate and myristate for ester and amide linkages, respectively, suggests the existence of at least two distinct classes of protein acyltransferases: A class that esterifies palmitate to cysteines and a class that attaches myristate to free amino groups on proteins. Currently, no protein acyltransferases have been isolated or localized intracellularly. The acyl donor and acceptor specificities of the enzymes involved in the acylation of proteins have also not been thoroughly examined. Using a cell-free system for acylation, Slomiany and coworkers (57, 58) have demonstrated a fatty acyltransferase activity that catalyzes the transfer of palmitic acid from palmitoyl CoA to deacylated mucus glycoprotein from gastric mucosa. Subcellular fractionation of rat gastric mucosa revealed that the highest specific activity of the enzyme was in the Golgi-rich fraction. Berger and Schmidt (3) have also demonstrated recently that palmitoyl CoA serves as the acyl donor for acylation of deacylated Semliki Forest virus *in vitro*. The significance of these studies is unclear, however, since the cell-free acylation exhibited no specificity, with respect to fatty acyl chain length or degree of saturation of the acyl chain used as substrate—in striking contrast to the remarkable degree of specificity observed in studies of acylation in BC3H-1 cells (40).

The majority of studies that have examined the biogenesis of acylproteins have focused on the viral membrane glycoproteins due to their high level of expression in virus-infected cells. Using virus-infected chicken embryo fibroblasts, Schmidt and Schlesinger (48) demonstrated that palmitate incorporation into Sindbis virus E2 and VSV-G glycoproteins continued for 10 to 20 minutes following inhibition of protein synthesis with cycloheximide, indicating that acylation of these glycoproteins is a relatively early posttranslational modification. To determine the approximate subcellular site of acylation, the timing of fatty acid incorporation into viral glycoproteins was compared to the processing of glycoprotein oligosaccharide side chains. These studies indicated that acylation of VSV-G protein occurs immediately prior to the completion of oligosaccharide trimming, which is known to occur within the Golgi apparatus. Dunphy et al. (14) and Quinn et al. (43) have also demonstrated that fatty acids are added to G glycoprotein in a smooth membrane fraction that contains oligosaccharide-trimming activity, suggesting that the acylating activity is located in the cis-Golgi or transitional elements of the endoplasmic reticulum (ER). These results are supported by the studies of Johnson and Schlesinger (22). Using the ionophore monensin, which allows transport of membrane proteins from the ER to the Golgi, but which blocks further transport to the cell surface, these investigators demonstrated that acylation of VSV-G occurred normally. In the presence of tunicamycin, however, nonglycosylated VSV-G protein does not move to the Golgi and is not acylated (48), providing additional evidence that acylation of viral envelope glycoproteins occurs early in the pathway of glycoprotein maturation.

Studies on the biosynthesis of normal cellular acylproteins have been limited primarily to the transferrin receptor (37, 38). In this case, acylation appears to be a late posttranslational event that occurs after oligosaccharide processing is complete. Unlike the viral glycoproteins, palmitate attachment to the receptor occurs normally in the presence of tunicamycin or cycloheximide. The possibility that the receptor also undergoes deacylation during its lifetime is suggested by pulse chase experiments, in which the fatty acid moiety on the receptor has been shown to turnover at a rate greater than that of receptor degradation. Thus, the kinetics of acylation of this cellular membrane glycoprotein clearly differ from those of the viral membrane glycoproteins. Further work is necessary in order to determine whether other cellular acylproteins undergo acylation (and possibly deacylation) by mechanisms similar to the transferrin receptor, or whether this glycoprotein is unique in its posttranslational processing.

Because acylproteins containing amide-linked fatty acids have only recently been identified, virtually no information is available regarding the biosynthesis of this class of acylproteins. The fact that many myristate-containing proteins are soluble (8, 34) and are synthesized on free polysomes (27) suggests that at least some myristoyl N-acyltransferase activity may be localized to the cytosol. The subcellular distribution of the enzymes involved in myristylation of proteins has not been examined, nor has the time of addition of this fatty acid to newly synthesized proteins been established. It also remains to be determined whether all of the membrane-associated proteins containing myristate are peripheral membrane proteins or whether some are also integral membrane glycoproteins. Because of the remarkable fatty acyl chain specificity of protein acylation, it seems reasonable to postulate that both the time of fatty acid addition as well as the subcellular site(s) of palmitate and myristate acylation may be very different.

Recent studies on myristate attachment to newly synthesized acylproteins in chicken embryo fibroblasts and BC3H-1 cells indicate that this is a very early covalent modification (8, 36). Simultaneous addition of cycloheximide and [3H]myristate to cells results in a complete block in acylation of all the major cellular acylproteins, indicating that myristate is added to newly synthesized acylproteins during, or very soon after, translation. In future studies, it will be important to determine more precisely as to when myristate is attached to acylproteins, and to investigate whether soluble and membrane-associated myristate-containing proteins are acylated with the same kinetics and by the same mechanisms.

Several characteristics of myristate-containing proteins suggest the possibility that the attachment of this fatty acid actually might occur cotranslationally. First, the labeling experiments described above indicate that myristylation occurs during or very soon after translation, which in the case of integral membrane proteins would restrict this modification to the ER or cis-Golgi complex. Second, the findings that many myristate-containing proteins are soluble (34) and are synthesized on free polysomes (27) indicate that myristylation does not require transit of newly synthesized acylproteins through a membrane compartment that contains the acylating enzymes. The myristylation of soluble proteins also suggests that at least some of the enzymes involved in myristate attachment to proteins may be localized to the cytosol, where they might have access to nascent polypeptide chains. Third, calcineurin b (1), the catalytic subunit of the cAMP-dependent protein kinase (9), p60^{src} (12), and p15 (20)—the only myristate-containing proteins to be studied in detail—have each been found to be acylated on an amino terminal glycine. Attachment of myristate to a glycine residue at the amino terminus requires the removal of the initiator methionine and possibly other amino acid residues. Cleavage of the initiator methionine is a common modification that has been shown to occur for a wide range of proteins after synthesis of the first 30 to 40 amino acids, while the nascent polypeptide is attached to the ribosome (40, 41). In a number of proteins, the new amino terminus is then rapidly acetylated. Attachment of acetate to the amino terminus of newly synthesized proteins has been shown to occur *in vivo* and *in vitro* by a ribosome-associated protein acetyltransferase that uses acetyl CoA as an acetyl donor (4, 60). It is tempting to speculate that a myristoyl N-acyltransferase and an acetyltransferase, both of which modify amino terminal amino acids, might have similarities in their mechanisms. If the myristoyl N-acyltransferase were associated with ribosomes in a manner similar to the acetyltransferases (60), it would provide a mechanism whereby one enzyme could acylate both soluble proteins and integral membrane glycoproteins. Obviously, a number of other possible mechanisms for myristylation can be postulated; however, the possibility that myristylation occurs cotranslationally is relatively easy to test; and if it is shown to be the mechanism for acylation, it would have broad implications for the role of this covalent modification. Moreover, if myristylation is shown to be cotranslational, this would represent an important first step toward the eventual purification of the protein acyltransferases, since one would know the approximate subcellular location of these enzymes.

A complete understanding of the biosynthetic events involved in the attachment of fatty acids to proteins will require the development of cell-free systems that exhibit the same acyl donor and acceptor specificities observed *in vivo*. The enzymology of palmitate incorporation into proteins is amenable to these types of studies, because palmitate can be easily removed from acylproteins by treatment with hydroxylamine and the deacylated proteins can then be used as substrates for acylation. The development of a similar system for *in vitro* acylation of proteins with myristate presents more difficulties, because there are no current methods for specifically releasing myristate from proteins. Therefore, the production of a substrate for *in vitro* myristylation will require the identification of the amino acid sequence that serves as the specific recognition signal for myristate attachment. Synthetic peptide substrates can then be prepared and used as artificial acceptors for myristylation *in vitro*. A similar approach has been used for studying N-linked glycosylation (61) and acetylation of proteins (18).

Based on the studies discussed above, a number of possible pathways for synthesis, processing, and intracellular transport of acylproteins can be envisioned. Some of these pathways are illustrated in Figure 6.2. It is clear that palmitate is esterified to a number of integral membrane glycoproteins as they are transported from the rough endoplasmic reticulum (RER) through the Golgi apparatus, en route to the cell surface. It is not yet known whether the enzymes involved in the attachment of palmitate to proteins are also located in other regions of the cell, or whether palmitate-containing proteins are found in membranes other than the plasma membrane. It is also well established that many myristate-containing proteins are synthesized and acylated in the cytosol. Can myristate also be added to integral membrane glycoproteins; and, if so, in what subcellular compartments do the myristoyl N-acyltransferases involved in this modification reside? Are myristylated proteins targeted to specific subcellular organelles or membrane systems? The answers to these questions must await further studies.

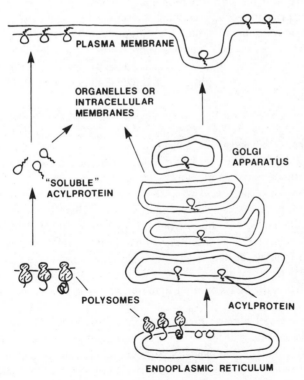

Fig. 6.2. Possible pathways for processing and intracellular transport of acylproteins.

Functions of Covalent Fatty Acid

Despite the wide range of fatty-acylated proteins that have been identified, the precise function of this covalent modification has not been determined and will no doubt be one of the most difficult questions to address in the study of protein acylation. The fact that many cell-surface proteins are not acylated indicates that covalent fatty acid is not required for intracellular transport to the plasma membrane. However, it is possible that covalent fatty acid might play a role in targeting acylproteins to specific subcellular membrane systems or organelles. The subcellular distribution of acylproteins has not yet been examined, and it will be an interesting subject for future studies.

Studies of the retrovirus-transforming proteins suggest that fatty acid may serve as an anchor for binding otherwise soluble proteins to membranes. Using deletion mutants in the N terminal region of p60^{src}, Cross et al. (12) demonstrated that the site of myristate attachment is contained in amino acids 1 to 14. Deletion of this region prevents myristylation. Under these conditions, p60^{src} does not associate with membranes and is transformation-defective. Similar studies have been carried out with p21^{ras}, which contains covalent palmitate near the C terminus (63). Mutants containing deletions located at or near the C terminus of p21 were not acylated and did not associate with membranes. These mutants were also transformation-defective. Thus, in the case of retrovirus-transforming proteins, acylation not only appears to be responsible for mediating the interaction of "soluble" proteins with membranes, but it also appears to be the key step required for cell transformation. These results should be interpreted with some caution, however, since it is unclear as to what effects these relatively large amino acid deletions might have on the secondary structure and/or function of the protein.

A role for covalent fatty acid in influencing the properties of the lipid bilayer surrounding acylproteins has been suggested by Petri et al. (42). These investigators fluorescently labeled VSV-G protein by growing virus-infected hamster kidney cells in the presence of 16(9-anthroyloxy)palmitate. Labeled G protein was then reconstituted into dipalmitoylphosphatidylcholine vesicles, and the mobility of protein-bound fatty acid was examined as a function of temperature by fluorescence measurements. These studies indicated that the covalent lipid moiety strongly interacted with the surrounding bilayer, resulting in the removal of phospholipid in the region of G protein from the phase transition. These data also demonstrate that the acylation site on G protein is located within a protein domain that interacts with the membrane, and that the covalent fatty acid is associated within the bilayer, rather than buried within the secondary structure of the polypeptide.

Mycoplasma capricolum, a prokaryotic sterol and fatty acid auxotroph, has been shown to contain a number of membrane proteins that are specifically acylated with palmitate and oleate (13). Acylation of the oleate-containing proteins is stimulated in the presence of cholesterol. This is the only example described to date in which acylation has been shown to be modulated, suggesting the interesting possibility that membrane fluidity may influence protein acylation.

Slomiany and coworkers (55–58) have examined the function of covalent fatty acid on the human gastric mucus glycoprotein. Purified mucus glycoprotein is resistant to digestion with pronase, whereas following removal of covalent fatty acid by treatment with hydroxylamine, the glycoprotein becomes susceptible to pronase. Mucus glycoprotein, isolated from patients with cystic fibrosis, contains two- to threefold more covalent fatty acid per molecule and is highly resistant to pronase. These data suggest that covalent fatty acid protects this acylprotein from proteolytic digestion, and that the increased acylation of mucus glycoprotein in cystic fibrosis may prevent normal turnover of this protein and contribute to the abnormal accumulation of poorly soluble secretions associated with the disease.

In an effort to determine the role of acylation in the expression of envelope virus glycoproteins, Schlesinger and Malfer (47) treated virus-infected chicken embryo fibroblasts with cerulenin (2,3 epoxy-4-oxo-7,10 dodecadienoylamide), which inhibits *de novo* fatty acid synthesis by binding irreversibly to the β-ketoacyl acyl carrier protein synthetase. In the presence of cerulenin, G glycoprotein from the Indiana serotype of vesicular stomatitis virus (VSV_{Ind}) was not acylated, and assembly and budding of virions was blocked. However, transport of VSV-G to the cell surface occurred normally in cells treated with the drug. Cerulenin also inhibits ACh receptor expression in BC3H-1 cells by blocking assembly of receptor subunits in a multisubunit receptor complex—a process that normally occurs in the Golgi apparatus (35). Together, these studies suggest that acylation might play a role in protein-protein or protein-lipid interactions within membranes. The mechanism by which cerulenin inhibits acylation of envelope virus glycoproteins with exogenous [³H]palmitate is unclear. One possibility is that cerulenin, which is a 12-carbon fatty acid amide, may act as a fatty acid analog blocking the active site of the acyltransferase involved in the acylation reaction.

In contrast to the apparent role of covalent fatty acid in the assembly of G protein from VSV_{Ind} into virions, studies by Gallione and Rose (16) demonstrate that acylation is not an absolute prerequisite for virion assembly. Using the New Jersey serotype of VSV—which contains a number of amino acid substitutions in the domain cytoplasmic tail domain, where the acylation site normally appears to reside—these investigators demonstrated that this G protein is nonacylated. However, virions are normally formed in cells infected with this strain of the virus. As in the studies of p60[src] described above, it is somewhat difficult to interpret results of studies such as these, in which proteins have undergone large changes in their primary structure. These changes in amino acid sequence could result in alterations in the secondary structure of the proteins that eliminate the need for a fatty acid moiety.

Currently, there are no inhibitors of fatty acylation of cellular proteins, which have been thoroughly characterized. The identification of specific inhibitors of N-linked and SH-linked acylation would be of major importance for an understanding of the function of this covalent modification. The development of temperature-sensitive mutants that exhibit defects specifically in acylation of proteins would also contribute to our knowledge of the significance of acylation.

Summary and Future Directions in the Study of Protein Acylation

Acylproteins represent a major class of cellular proteins whose identities and functions are only beginning to be understood. As was found for glycosylation, multiple functions as well as biosynthetic pathways for fatty acylation will probably be discovered in the future. From the limited work that has been done on protein acylation, a number of important questions regarding both the role of fatty acid acylation and the biosynthetic events involved in this covalent modification of protein can now begin to be addressed. For example, when are the palmitate and myristate attached to newly synthesized acylproteins (*i.e.*, cotranslationally or posttranslationally)? And where are the protein acyltransferases localized within the cell? Are membrane-bound acylproteins localized to specific membrane systems? And, if so, how do these proteins make their way from the site of synthesis to their destination within a specific membrane system? What is the function of the fatty acid moiety on acylproteins? What are the characteristics of a protein that specify whether or not it becomes acylated? What are the identities of the amino acid residues that serve as acylation sites? With the continued identification of cellular acylproteins and the development of *in vitro* systems for studying the biosynthetic steps involved in the attachment of fatty acids to proteins, many of these questions will be answered in the near future.

References

1. Aitken, A., Cohen, P., Santikarn, S., Williams, D.H., Calder, A.G., Smith, A. & Klee, C.B. (1982) FEBS Lett. *150*, 314–318.
2. Agrawal, H.C., Randle, C.L. & Agrawal, D. (1982) J. Biol. Chem. *257*, 4588–4592.
3. Berger, M. & Schmidt, M.F.G. (1984) J. Biol. Chem. *259*, 7245–7252.
4. Bloamendal, H. (1977) Science *197*, 127–138.
5. Bolanowski, M.A., Earles, B.J. & Lennarz, W.J. (1984) J. Biol. Chem. *259*, 4934–4940.
6. Braun, V. & Rehn, K. (1969) Eur. J. Biochem. *10*, 426–438.
7. Brugge, J., Yonemoto, W. & Darrow, D. (1983) Mol. Cell Biol. *3*, 9–19.
8. Buss, J.E., Kamps, M.P. & Sefton, B.M. (1984) Mol. Cell Biol. *4*, 2697–2704.
9. Carr, S.A., Biemann, M., Shoji, S., Parmelee, D.C. & Titani, K. (1982) Proc. Natl. Acad. Sci. USA *80*, 339–343.
10. Conti-Tronconi, B., Gotti, C.M., Hunkapiller, M.W. & Raftery, M.A. (1982) Science *218*, 1227–1229.
11. Courtneidge, S.A. & Bishop, J.M. (1982) Proc. Natl. Acad. Sci. USA *79*, 7117–7112.

12. Cross, F.R., Garber, E.A., Pellman, D. & Hanafusa, H. (1984) Mol. Cell Biol. *4*, 1834–1842.
13. Dahl, C.E., Dahl, J.S. & Block, K. (1983) J. Biol. Chem. *258*, 11814–11818.
14. Dunphy, W.G., Fries, E., Urbani, L.J. & Rothman, J.E. (1981) Proc. Natl. Acad. Sci. USA *78*, 7453–7457.
15. Folch-Pi, J. & Lees, M.B. (1951) J. Biol. Chem. *191*, 807–817.
16. Gallione, C.J. & Rose, J.K. (1983) J. Virol. *46*, 162–169.
17. Garber, E.A., Krueger, J.G., Hanafusa, H. & Goldberg, A.R. (1983) Nature *302*, 161–162.
18. Granger, M., Tesser, G.I., DeJong, W.W. & Blamendal, H. (1976) Proc. Natl. Acad. Sci. USA *73*, 3010–3014.
19. Hantke, K. & Braun, V. (1973) Eur. J. Biochem. *34*, 284–296.
20. Henderson, L.E., Krutzsch, H.C. & Oroszlan, S. (1983) Proc. Natl. Acad. Sci. USA *80*, 339–343.
21. Henning, R. & Lang-Mutschler, J. (1983) Nature *305*, 736–738.
22. Johnson, D.C. & Schlesinger, M.J. (1980) Virology *103*, 407–424.
23. Kaufman, J.F., Krangel, M.S. & Stominger, J.L. (1984) J. Biol. Chem. *259*, 7230–7238.
24. Keenan, T.W., Heid, H.W., Stadler, J., Jarasch, E.-D. & Franke, W.W. (1982) Eur. J. Cell Biol. *26*, 270–276.
25. Khandwala, A.S. & Kasper, C.B. (1971) J. Biol. Chem. *246*, 6242–6246.
26. Klenk, H.-D. & Choppin, P.W. (1969) Virology *38*, 255–268.
27. Levinson, A.D., Courtneidge, S.A. & Bishop, J.M. (1980) Proc. Natl. Acad. Sci. USA *78*, 1624–1628.
28. MacLennon, D.H., Yip, C.C., Iles, G.H. & Seeman, P. (1972) Cold Spring Harbor Symp. Quant. Biol. *37*, 469–477.
29. Magee, A.I., Koyama, A.H., Malfer, C., Wen, D. & Schlesinger, M.J. (1984) Biochim. Biophys. Acta *798*, 156–166.
30. Magee, A.I. & Schlesinger, M.J. (1982) Biochim. Biophys. Acta *694*, 279–289.
31. Marinetti, G.V. & Cattieu, K. (1982) Biochim. Biophys. Acta *685*, 109–116.
32. Neimann, H. & Klenk, H.D. (1981) J. Mol. Biol. *153*, 993–1010.
33. O'Brien, P.J. & Zatz, M. (1984) J. Biol. Chem. *259*, 5054–5057.
34. Olson, E.N., Towler, D.A. & Glaser, L. (1985) J. Biol. Chem. *260*, 3784–3790.
35. Olson, E.N., Glaser, L. & Merlie, J.P. (1984) J. Biol. Chem. *258*, 5364–5367.
36. Olson, E.N. & Spizz, G. (1986) J. Biol. Chem. *261*, 2458–2466.
37. Omary, M.B. & Trowbridge, I.S. (1981) J. Biol. Chem. *256*, 4715–4718.
38. Omary, M.B. & Trowbridge, I.S. (1981) J. Biol. Chem. *256*, 12888–12892.
39. Ozols, J., Carr, S.A. & Strittmatter, P. (1984) J. Biol. Chem. *259*, 13349–13354.
40. Palmiter, R.D., Gagnon, J. & Walsh, K.A. (1978) Proc. Natl. Acad. Sci. USA *75*, 94–98.
41. Palmiter, R.D. (1977) J. Biol. Chem. *252*, 8781–8783.
42. Petri, W.A., Pal, R., Barenholz, Y. & Wagner, R.R. (1981) J. Biol. Chem. *256*, 2625–2627.
43. Quinn, P., Griffiths, G. & Warren, G. (1983) J. Cell Biol. *96*, 851–856.
44. Rose, J.K., Welch, W.J., Sefton, B.M., Esch, F.S. & Ling, N.C. (1980) Proc. Natl. Acad. Sci. USA *77*, 3884–3888.
45. Rose, J.K., Adams, G.A. & Gallione, C.J. (1984) Proc. Natl. Acad. Sci. USA *81*, 2050–2054.

46. Schlesinger, M.J., Magee, A.I. & Schmidt, M.F.G. (1980) J. Biol. Chem. *255*, 10021–10024.
47. Schlesinger, M.J. & Malfer, C. (1982) J. Biol. Chem. *257*, 9887–9890.
48. Schmidt, M.F.G. & Schlesinger, M.J. (1979) Cell *17*, 813–819.
49. Schmidt, M.F.G., Bracha, M. & Schlesinger, M.J. (1979) Proc. Natl. Acad. Sci. USA *76*, 1687–1691.
50. Schmidt, M.F.G. & Schlesinger, M.J. (1980) J. Biol. Chem. *259*, 3334–3339.
51. Schmidt, M.F.G. (1983) Curr. Topics Micro. Immunol. *102*, 101–124.
52. Schmidt, M.F.G. (1982) Virology *116*, 327–338.
53. Schultz, A.M., Henderson, L.E., Oroszlan, S., Garber, E.A. & Hanafusa, H. (1985) Science *227*, 427–429.
54. Sefton, B.M., Trowbridge, I.S. & Cooper, J.A. (1982) Cell *31*, 465–474.
55. Slomiany, A., Witas, H., Aono, M. & Slomiany, B.L. (1983) J. Biol. Chem. *258*, 8535–8538.
56. Slomiany, A., Slomiany, B.L., Witas, H., Aono, M. & Newman, L.J. (1983) Biochem. Biophys. Res. Comm. *113*, 286–293.
57. Slomiany, A., Jozwiak, Z., Takogi, A. & Slomiany, B.L. (1984) Arch. Biochem. Biophys. *229*, 560–567.
58. Slomiany, A., Liau, Y.H., Takogi, A., Laszewica, W. & Slomiany, B.L. (1984) J. Biol. Chem. *259*, 13304–13308.
59. Stoffel, W., Hillen, H., Schroeder, W. & Deutzmann, R. (1983) Hoppe-Seyler's Z. Physiol. Chem. *364*, 1344–1366.
60. Traugh, J.A. & Sharp, S.B. (1977) J. Biol. Chem. *252*, 3738–3744.
61. Welply, J.K., Shenbagamurthi, P., Lennarz, W.J. & Naider, F. (1983) J. Biol. Chem. *258*, 11856–11863.
62. Wen, D. & Schlesinger, M.J. (1984) Mol. Cell Biol. *4*, 688–694.
63. Williamsen, B.M., Christensen, A., Hubbert, N.L., Papageorge, A.G. & Lowy, D.R. (1984) Nature *310*, 583–586.

7

Biosynthesis of Endoplasmic Reticulum Membrane Proteins

CECIL B. PICKETT and CLAUDIA A. TELAKOWSKI-HOPKINS

The intent of this review is to discuss recent literature on the biosynthesis of endoplasmic reticulum (ER) membrane proteins, with a focus on two mechnisms that appear to be operative during insertion of these proteins into the ER membrane (*i.e.*, cotranslational and posttranslational insertion). During the past few years, the biosynthesis of membrane proteins in general has received much attention, particularly in regard to the events involved in the synthesis, processing, transfer, and insertion of proteins into or across their appropriate membrane (for reviews, see [1–5]). Although a significant literature has accumulated on the biosynthesis of secretory proteins, only a limited number of studies to date have focused on the biosynthesis of ER membrane proteins. This has been due, in part, to the inherent difficulties in purifying to homogeneity many intrinsic ER membrane proteins.

Turnover of Endoplasmic Reticulum Membrane Proteins

The ER represents a major cellular constituent that is comprised of an extensive network of tubules, vesicles, and lamellae. It is highly developed in those cell types that are actively involved in protein synthesis and secretion (*e.g.*, liver and exocrine cells). The ER is a dynamic structure that is adaptive to various physiologic and pharmacologic stimuli, such as diet, starvation, various hormones, and xenobiotics. A group of ER membrane proteins whose biosynthesis has been the subject of a number of investigations are the drug-metabolizing enzymes, which include the cytochrome P-450 isozymes, epoxide hydrolase, NADPH-cytochrome P-450 reductase, cytochrome b_5, NADH-cytochrome b_5 reductase, and UDP-glucuronosyltransferases. These enzymes function in the metabolism of a variety of drugs, mutagens, and carcinogens as well as in endogenous compounds such as fatty acids and various steroids. The first studies on the biosynthesis of ER membrane proteins focused primarily on the turnover of NADPH-cytochrome c (P-450) reductase and cytochrome b_5 *in vivo*. In 1967, Omura *et al.* (6), using ^{14}C-leucine as the radiolabeled amino acid, reported that total ER membrane proteins have a half-life of 75 to 113 hours, while the half-life of NADPH-cytochrome c reductase and cytochrome b_5 were 80 and 120 hours, respectively. Utilizing a double-labeling technique, Arias *et al.* (7) confirmed the differences in half-life for NADPH-cytochrome c reductase and cytochrome b_5, but found that when ^{14}C-quanidino-L-arginine was used as the radiolabeled precursor instead of ^{14}C-arginine, the half-life of total microsomal protein was 48 hours compared to 5 days. This striking difference in half-life of total microsomal protein was due to differences in the re-utilization of the two labeled precursors, and it emphasized the need to use radiolabeled precursors that were not reutilized to significant extents. In order to eliminate the influence of amino acid reutilization on turnover rates, Swick and Ip (8) recommended the use of $NaH^{14}CO_3$ to label arginine *in vivo*.

Recently, Parkinson *et al.* (9), using immunoprecipitation techniques, determined the *in vivo* turnover rates of liver microsomal epoxide hydrolase and both the heme and apoprotein moieties of cytochrome P-450a, P-450b+P-450e, and P-450c in Aroclor 1254-pretreated rats by following the decay in specific radioactivity from 2 to 96 hours after simultaneous injection of $NaH^{14}CO_3$ and ^{3}H-labeled δ-amino-levulinic acid. Cytochrome P-450b and P-450e are the major cytochrome P-450s induced by phenobarbital and show immunologic identity (10, 11). Cytochrome P-450c is a major isozyme induced by 3-methylcholanthrene (10). Cytochrome P-450a has been purified from PB-, 3MC-, and Aroclor-treated rats, and it preferentially catalyzes the hydroxylation of testosterone at the 7α position (10). Parkinson *et al.* (9) found that cytochrome P-450a displayed a biphasic turnover of both heme and apoprotein moieties of the holoenzyme. The half-lives of the apoprotein were 12 and 52 hours for the fast- and slow-phase components, respectively, and 5 and 34 hours for the heme moiety. Similarly, total microsomal protein also displayed a biphasic turnover of 5 to 9 hours for the fast phase and 82 hours for the slow phase. Total heme in these membranes had a half-life of 8 hours for the slow phase and 41 hours for the fast phase. Cytochrome P-450b+P-450e and cytochrome P-450c had half-lives of 37 and 28 hours for apoprotein and heme, respectively. In contrast, epoxide hydrolase, which converts epoxides generated by cytochrome P-450 into transdihydrodiols, had a half-life of 132 hours, which was significantly longer than the half-life of the slow phase of the average microsomal protein. Sadano and Omura (12) have also measured the turnover rates using [^{14}C]-NaHCO$_3$ of cytochrome P-450 (PB), cytochrome P-450 (MC), and cytochrome b_5 in normal rats and rats pretreated with phenobarbital. These investigators found that the apoprotein moieties of cytochrome P-450 (PB), cytochrome P-450 (3MC), cytochrome b_5, and NADPH-cytochrome P-450 reductase have half-lives of 25, 15, 50, and 35 hours, respectively, in normal rats compared to 20, 20, 30, and 25 hours, respectively, in phenobarbital-treated rats. In contrast, however, the heme moiety in cytochrome P-450 (PB) turned over with a half-life of 15 hours in normal rats, indicating that the heme and apoprotein moieties of cytochrome P-450 turn over asynchronously.

Very recently, Shiraki and Guengerich (13) measured the turnover of seven rat liver cytochrome P-450 isozymes, NADPH-cytochrome P-450 reductase, and epoxide hydrolase in untreated, phenobarbital-treated, and β-napthoflavone-treated rats. In general, the half-lives of these proteins in untreated rats were all very similar ($\sim$ 10 to 29 hours), and neither phenobarbital nor β-napthoflavone markedly altered these rates. Furthermore, the data obtained by Shiraki and Guengerich are in general agreement with turnover rates reported previously by Parkinson et al. (9) and Sadano and Omura (12). However, Shiraki and Guengerich (13) found that the half-life of epoxide hydrolase was 24 hours after phenobarbital administration, whereas Parkinson et al. reported a half-life of 132 hours for epoxide hydrolase after Aroclor 1254 treatment. The differences in the turnover rate of epoxide hydrolase reported by Parkinson et al. (9) and Shiraki and Guengerich (13) may be due, in part, to the use of rats that were pretreated with two different xenobiotics; namely, phenobarbital and Aroclor 1254.

In contrast to the turnover rates of the drug-metabolizing enzymes is the rapid turnover rate reported for HMG-CoA reductase, which is a membrane protein of the ER that catalyzes the rate-limiting step in sterol biosynthesis. Edwards and Gould (14) measured the turnover rate of HMG-CoA reductase after inhibiting protein synthesis by cycloheximide administration and found that the reductase had a half-life of 4.2 hours. More recently, Edwards et al. (15), using specific immunoprecipitation techniques, reported the half-life of HMG-CoA reductase in isolated hepatocytes to be 114 minutes. Interestingly, the administration of mevalonolactone to rats reduced the half-life to 37 minutes by increasing the rate of degradation of HMG-CoA reductase three-fold. Based upon the differential turnover rates of various microsomal proteins, it is now generally accepted that ER membrane proteins are not synthesized and degraded as a unit.

Mechanisms of Insertion of Proteins Into the Endoplasmic Reticulum Membrane

Although only a limited number of studies have been done to elucidate mechanisms by which microsomal proteins are incorporated into the ER membrane, it appears that two distinct mechanisms are responsible for their insertion. The first mechanism involves cotranslational insertion of the polypeptide, which is mediated by the signal recognition particle (SRP) and the SRP receptor (2, 5). The second mechanism, posttranslational insertion, is not protein mediated, but does require the presence of an insertion sequence in the protein (1). Examples of microsomal proteins that are incorporated into the microsomal membrane either by cotranslational or posttranslational insertion are presented below.

Cotranslational Insertion of Proteins Into the Endoplasmic Reticulum Membrane

Early studies on the biosynthesis of phenobarbital-inducible cytochrome P-450 focused on elucidating the site of synthesis of cytochrome P-450 in the cell. Negishi *et al.* (16) demonstrated that I^{125}-labeled antibodies to cytochrome P-450 bound preferentially to membrane-bound polysomes, rather than to free polysomes, and that nascent cytochrome P-450 polypeptides were found in membrane-bound polysomes. Later, Takagi (17) and Ohlsson and Jergil (18) confirmed the observations by Negishi *et al.* (16).

In 1979, Dubois and Waterman (19) used *in vitro* translation analysis and specific immunoprecipitation to demonstrate that the primary translation product of the major phenobarbital-inducible cytochrome P-450 comigrated on one-dimensional SDS-polyacrylamide gels with the mature enzyme. These data provided the first indication that the incorporation of cytochrome P-450 into the ER membrane may not require the prescence of a cleavable signal peptide, which had been demonstrated to be a common feature of secretory proteins. Bar-Nun *et al.* (20) confirmed the absence of a cleavable NH_2 terminal signal sequence by demonstrating that the sequence of the primary translation product synthesized in the rabbit reticulocyte cell-free system was identical to the NH_2 terminal sequence of purified cytochrome P-450. The absence of any significant post-translational processing of the phenobarbital-inducible cytochrome P-450s has also been confirmed by two-dimensional (2D) gel electrophoresis (*i.e.*, isoelectric focusing in the first dimension followed by SDS-polyacrylamide gel electrophoresis in the second). Walz *et al.* (21) and Pickett *et al.* (22) demonstrated that the major phenobarbital-inducible cytochrome P-450s comigrated after 2D gel electrophoresis with *in vitro*-synthesized cytochrome P-450s. More recently, cDNA clones complementary to P-450b and P-450e mRNA have been sequenced, and the deduced NH_2 terminal amino acid sequences of the P-450s were in total agreement with the NH_2 terminal sequences of the purified cytochrome P-450s (23, 24). These studies established definitively that the phenobarbital-inducible cytochrome P-450s do not have NH_2 terminal signal sequences that are removed proteolytically.

In contrast to the phenobarbital-inducible cytochrome P-450, Kumar and Padmanaban (25) reported that the major 3-methylcholanthrene-inducible isozyme of cytochrome P-450 was synthesized as a higher molecular weight precursor (molecular weight=59,000) that was converted to the mature form (M_r, 53,000) by rat liver microsomal membranes. However, later studies (26–28), also using *in vitro* translation analysis, demonstrated that the primary translation product of this isozyme of cytochrome P-450 was identical in size to the purified P-450. The discrepancy between Kumar and Padmanaban's results and data obtained by other investigators is unclear, but it may be due to the specificity of the antiserum used in the studies. Recently, cDNA clones complementary to the rat liver 3-methylcholanthrene-inducible P-450s have been constructed (29, 30). The amino acid sequence deduced from DNA sequence analysis of these cDNAs indicates that the primary translation product most likely has an NH_2 terminal sequence identical to the purified isozymes.

Two other ER membrane proteins involved in drug metabolism whose biosynthesis have been examined are epoxide hydrolase (molecular weight=52,000) and NADPH-cytochrome P-450 reductase (molecular weight=77,000). When poly(A^+)-RNA was isolated from untreated or phenobarbital-treated rats and translated *in vitro*, the immunoprecipitated primary-translated product of epoxide hydrolase comigrated on SDS-polyacrylamide gels with the mature enzyme (31–34). Gonzalez and Kasper (31) confirmed that the *in vitro*-synthesized enzyme was virtually identical to the mature enzyme by limited peptide mapping using *Staphylococcus aureus* V-8 protease. These data indicated that like the cytochrome P-450s, epoxide hydrolase was not synthesized as a higher molecular weight precursor that was processed proteolytically during insertion into the ER membrane. Okada *et al.* (34) confirmed by sequence analysis that the amino terminal sequence of purified epoxide hydrolase corresponded to the NH_2 terminal sequence of epoxide hydrolase immunoprecipitated from the translation system. Similarly, Okada *et al.* (34) determined the partial amino acid sequence of *in vitro*-synthesized rat liver NADPH-cytochrome P-450 reductase, which is an enzyme required for electron transfer to cytochrome P-450 in various tissues; they found that among the first 15 amino acids, seven were hydrophobic amino acids (*i.e.*, leucine). Okada *et al.* (34) concluded that the amino terminal portion of the polypeptide represents a signal sequence that is responsible for insertion of the proteins into the ER and is important for anchoring it there.

Utilizing trypsin to solubilize NADPH-cytochrome P-450 reductase from the ER membrane, Black and Coon (35) have isolated the hydrophobic membrane domain of rabbit and rat liver NADPH-cytochrome P-450 reductase. In contrast to the rat sequence determined by Okada *et al.* (34), Black and Coon (35) found no leucine residues in the first 15 amino acids of the rabbit or rat liver NH_2 terminal sequence. More recently, Porter and Kasper (36) have determined the DNA sequence of a rat liver NADPH-cytochrome P-450 reductase cDNA clone. The deduced NH_2 terminal amino acid sequence is in agreement with the sequence data of Black and Coon (35). However, sequence analysis of the rabbit and rat NH_2 terminal domain reveals a very hydrophobic region involved in the binding of cytochrome P-450 and phospholipid (35). The degree of hydrophobicity of the NH_2 terminal domain of the reductase is consistent with the hydrophobicity of the NH_2 terminal regions of other ER membrane proteins.

Recently, Sakaguchi *et al.* (37) determined whether the SRP is required for the cotranslational insertion of cytochrome P-450 into microsomal membranes. The SRP is required for the cotranslational insertion of secretory proteins into rough microsomes. The SRP is not only required for the translocation of secretory proteins, but is also required for cotranslational insertion of the δ-subunit of Torpedo acetylcholine (ACh) receptor (a plasma membrane protein) into microsomes (38) as well as the translocation of the lysosomal protease cathepsin D across the microsomal protein (39). Sakaguchi *et al.* (37), using a wheat germ *in vitro* translation system supplemented with dog pancreatic microsomes, found that the insertion of the major phenobarbital-inducible cytochrome P-450 into salt-washed microsomes (microsomes minus SRP) was greatly diminished, as compared to unwashed rough microsomes. The insertion of cytochrome P-450 could be restored by supplementing the system with purified SRP. In the absence of microsomes, SRP inhibited the translation of the mRNA encoding the cytochrome P-450; however, the translation arrest could be released by the addition of salt-washed microsomes. These data definitely demonstrated the requirements of SRP for the insertion of P-450 into the membrane.

Similar to the cytochrome P-450 isozymes, the sarcoplasmic reticulum Ca^{2+} ATPase—a 119,000 dalton polypeptide from skeletal muscle—does not possess a cleavable NH_2 terminal signal sequence (40). In order for integration of the ATPase into the membrane to occur, the cotranslational presence of microsomal membranes is required, and insertion of the protein into the membrane is mediated by the SRP (40). However, unlike cytochrome P-450, there is no SRP-mediated elongation arrest of translation of the Ca^{2+} ATPase.

Brown and Simoni (41) have used a cell line C-100 derived from baby hamster kidney cells to study the biogenesis of HMG-CoA reductase, which is an integral membrane protein of the ER. This cell line is resistant to 225 μM compactin—a specific inhibitor of HMG-CoA reductase—and overproduces HMG-CoA reductase 100 fold. Brown and Simoni (41) have found that HMG-CoA reductase is synthesized on membrane-bound polysomes—not free polysomes—and inserts into dog pancreatic microsomes when these membranes are added cotranslationally, but not posttranslationally. When microsomes were heated at 60°C for 10 minutes to inactivate the translocation system, HMG-CoA reductase did not insert into the membrane. If salt-washed dog pancreatic membranes were added to the translation system, most of the HMG-CoA reductase remained in soluble form; whereas with the addition of purified SRP, 50% of the total HGM-CoA reductase was associated with the microsomal membranes. Therefore, analogous to cytochrome P-450 and the Ca^{2+} ATPase, it is apparent that the SRP is required for the insertion of the reductase into the membrane. The partial NH_2 terminal amino acid sequences of HMG-CoA reductase radiolabeled *in vivo* and *in vitro* were identical, suggesting that like the other ER membrane proteins, HMG-CoA reductase also does not possess a cleavable signal sequence.

Chin *et al.* (42) have recently constructed several overlapping cDNA clones complementary to hamster HMG-CoA reductase mRNA. Based upon nucleotide sequence analysis, these cDNAs encompass the entire coding region and much of the 5'- and 3'-untranslated regions of the mRNA. The NH_2 terminal amino acid sequence deduced from DNA sequence analysis is in total agreement with the NH_2 terminal amino acid sequence of the reductase labeled *in vivo* by ^{35}S-methionine and ^{3}H-phenylalanine. The data obtained by Chin *et al.* (42), as well as Brown and Simoni (41), demonstrate that the reductase does not have a cleavable NH_2 terminal sequence. The NH_2 terminal amino acids of various drug-metabolizing enzymes are presented in Figure 7.1 for comparison. As can be seen in this figure, all of these ER membrane proteins have NH_2 termini that are very hydrophobic.

Fig. 7.1. NH$_2$ terminal sequences of cytochrome P-450 isozymes, epoxide hydrolase, and HMG-CoA reductase. All the NH$_2$ terminal sequences of the rat liver cytochrome P-450 isozymes were taken from (67). (References are indicated in parentheses).

In contrast to the cytochrome P-450 isozymes (43), however, HMG-CoA reductase is glycosylated with an oligosaccharide of the "high-mannose" type. Liscum *et al.* (44) have utilized a cultured hamster cell line, UT-1, that grows in the presence of high concentrations of compactin. In this cell line, the reductase gene is amplified 15 fold, and approximately 500 times the normal amount of reductase is expressed. Liscum *et al.* (44) found that solubilized HMG-CoA reductase from the UT-1 cells, Chinese hamster ovary cells, and rat liver absorbed quantitively and specifically to concanavalin A-sepharose. Further-more, it was demonstrated that UT-1 cells incorporated [1,6-^{3}H-glucosamine] into reductase that upon digestion with endo-N-acetylglucosaminidase H, the radioactivity could be localized in N-linked high-mannose chains. The carbohy-drate of the HMG-CoA reductase is found on a 30,000 to 35,000-dalton fragment of the enzyme, which can be separated proteolytically from a cytoplasmic 53,000-dalton fragment containing the active site of the enzyme. These data suggest that HMG-CoA reductase spans the ER membrane with a portion of the polypeptide facing the lumen of the ER. In analyzing the protein sequence deduced from DNA sequence analysis of cDNA clones complementary to HMG-CoA reductase, Chin *et al.* (42) found three potential sites for asparagine-linked glycosylation.

Models of the orientation of HMG-CoA reductase by Liscum *et al.* (45) and cytochrome P-450 LM$_2$ by Black and Coon (personal communication) in the ER membrane are presented in Figures 7.2 and 7.3, respectively. Interestingly, these models predict that HMG-CoA reductase and cytochrome P-450 LM$_2$ traverse the membrane several times.

Although it would appear from the above discussion that a common feature of proteins inserted into the ER membrane cotranslationally is the lack of a cleav-able NH$_2$ terminal sequence, a recent report by Mackenzie and Owens (46) sug-

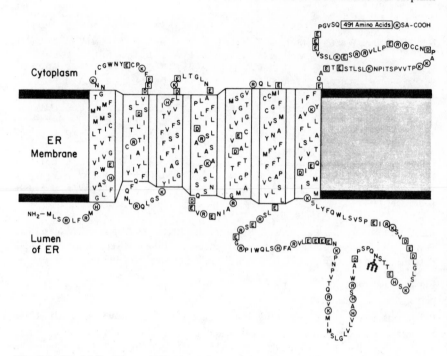

Fig. 7.2. Model for the possible orientation of the HMG-CoA reductase in the ER membrane. Amino acid residues are shown in the single-letter code. Amino acids with positively charged side chains are shown in circles. Amino acids with negatively charged side chains are shown in squares. The N-linked carbohydrate chain at residue 281 is indicated between the sixth and seventh membrane-spanning region. (This figure has been reprinted from [45] with permission of the authors and J. Biol. Chem.).

gests that an isozyme of the rat liver UDP-glucuronosyltransferases may be proteolytically processed by dog pancreatic microsomes. These transferases are a family of ER membrane proteins that function in the glucuronidation of many endogenous and exogenous lipophilic compounds. Mackenzie and Owens (46) observed that if poly(A+)RNA from rat liver was translated in a micrococcal nuclease-treated rabbit reticulocyte lysate system in the absence of dog pancreatic membranes, a polypeptide with a molecular weight of 52,000 could be immunoprecipitated with polyclonal antibodies raised against a mouse transferase isozyme. However, upon the inclusion of dog pancreatic membranes during *in vitro* synthesis, the 52,000-dalton form was converted to a polypeptide of 50,000 daltons, which could not be recognized by the antibody unless the membranes were solubilized first with detergent. These latter data suggest that the 50,000-dalton protein was incorporated into the microsomal membrane. Although these results provide the first example of an ER membrane protein that may be proteolytically processed cotranslationally, there are no sequence data

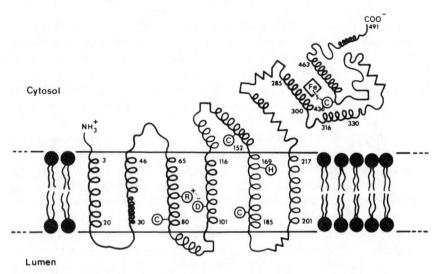

Fig. 7.3. Proposed topology of rabbit P-450 LM$_2$ in the ER membrane (Black and Coon, personal communication). Wide and narrow spiral segments indicate regions with predicted α and 3_{10} helices, respectively. Predicted β turns ([) and regions of β sheet (zigzag lines) are also shown.

yet on any purified transferase isozyme or the presumptive precursor to confirm the presence of a cleavable NH$_2$ terminal sequence.

Posttranslational Insertion of Proteins Into the ER Membrane

Liver microsomal cytochrome b$_5$ is an integral membrane protein (molecular weight = 15,223) that functions as a component of the microsomal electron transport chain (47–49). It is comprised of two domains: 1) a hydrophilic domain exposed to the aqueous environment and comprising 70% of the total protein; and 2) a hydrophobic domain comprised of ∼30 amino acids proximal to the COOH terminus. Purified cytochrome b$_5$ has been shown to bind to liposomes and microsomes *in vitro*, and binding is dependent upon the presence of the carboxyl terminal end (50). Cytochrome b$_5$ can also bind *in vitro* to peroxisomes, the outer mitochondrial membrane, the inner mitochondrial membrane, and Golgi (51).

NADH-cytochrome b$_5$ reductase is a microsomal flavoprotein (molecular weight = 35,000) consisting of a hydrophilic moiety containing an FAD prosthetic group and a hydrophobic segment, which is required for the binding of the reductase to microsomal as well as liposomal membranes and for the reconstitution of NADH-cytochrome c reductase activity in the presence of cytochrome b$_5$ (52–56). Mihara *et al.* (57) demonstrated that digestion of purified rabbit liver reductase with carboxypeptidase Y, but not aminopeptidase resulted in the removal of ~ 30 amino acid residues, 70% of which were hydrophobic. Concomitant with the removal of the hydrophobic carboxyl terminus was the loss of membrane binding and the ability to reconstitute NADH-cytochrome c reductase activity. Therefore, like cytochrome b_5, the binding of NADH-cytochrome b_5 reductase to microsomal membranes was thought to be mediated via a carboxyl terminal hydrophobic region. However, Kensil et al. (58) demonstrated that the membrane-binding domain of steer liver cytochrome b_5 reductase resides in the NH_2 terminal domain of that protein. These data are completely opposed to the results obtained by Mihara et al. (57). As described previously, these investigators found that the membrane-binding domain of rabbit liver cytochrome b_5 reductase was localized near the carboxyl terminal end. Although a species difference in the orientation of reductase in the membrane cannot be ruled out, the data presented by Kensil et al. (58) provide compelling evidence for localization of the membrane-binding domain in the NH_2 terminal region of the reductase. More recently, Ozols et al. (59) determined the complete amino acid sequence of the membrane-binding domain of cytochrome b_5 reductase, which is hydrophobic enough to serve as an insertion sequence. In addition, these investigators also found a myristic acyl chain on the NH_2 terminal segment. Although the myristic acyl chain is probably not essential for anchoring the reductase to the membrane, Ozols et al. (59) postulate that it may selectively stabilize a particular membrane structure and orientation that facilitates electron transport on the cytosolic surface of the ER.

Although early studies elucidating the site of synthesis of cytochrome b_5 were conflicting (60, 61), it was demonstrated later using both in vitro translation analysis and specific immunoprecipitation analysis that cytochrome b_5 as well as NADH:cytochrome b_5 reductase were synthesized by free polysomes, rather than by membrane-bound polysomes (62, 63). These findings are consistent with the idea that insertion of these proteins into microsomal membranes occurs posttranslationally. To test this hypothesis, Okada et al. (34) programmed the rabbit reticulocyte cell-free translation system with total rat liver in mRNA for 60 minutes, and aliquots of the translation mixture were incubated for 1 hour at room temperature with rough microsomal membranes stripped of ribosomes. After the incubation, samples were adjusted to 0.5 M NaCl and centrifuged for 30 minutes at 150,000×g. The resultant supernatant and microsomal pellets were subjected to immunoprecipitation using antiserum against cytochrome b_5 reductase or cytochrome b_5. Both cytochrome b_5 reductase and cytochrome b_5 were specifically localized in the pellet rather than the supernatant. These data indicated that both proteins were associated with the ER membrane after their synthesis was completed on free polysomes. Borgese and Gaetani (64) used carboxypeptidase Y as a proteolytic probe to assess whether the insertion of the NADH-cytochrome b_5 reductase after synthesis in vitro was identical to insertion of this protein in vivo. Microsomal membranes were incubated with carboxypeptidase Y in the presence and absence of Triton X-100, and were then analyzed for the presence of intact reductase by Western blotting. The native reductase is converted to a smaller form by the carboxypeptidase Y only when the microsomal membrane was disrupted with Triton X-100. Borgese and Gaetani (64) found that if reductase was synthesized in vitro in the absence of microsomes, carboxypeptidase Y converted the reductase to the smaller form. However, if the in vitro translation system was incubated with microsomes before addition of carboxypeptidase Y, approximately 70% of the reductase was protected from proteolysis in the presence of a sublytic concentration ($\sim 0.01\%$) of Triton X-100. If lytic concentrations of Triton X-100 were used and the integrity of the membrane was destroyed, carboxypeptidase Y converted the in vitro-synthesized reductase to the smaller form. These data very clearly illustrated that the reductase was inserting into the microsomal membranes posttranslationally in a manner analogous to what occurs in vivo.

Additional data supporting the idea that the incorporation of cytochrome b_5 into microsomal membranes was independent of cotranslational insertion of secretory proteins into microsomal membranes came from experiments by Bendzko et al. (65). These investigators found that when an excess of cytochrome b_5 was added to dog pancreatic microsomal membranes, the cotranslational processing of carp pancreatic preproinsulin to proinsulin by these membranes was unaffected. Consequently, there appears to be no competition between the insertion of b_5 with the cotranslational insertion of secretory proteins, indicating two distinct mechanisms are responsible for their insertion.

More recently, Anderson et al. (40) have found that the SRP is not required for insertion of in vitro-synthesized cytochrome b_5 into the dog pancreas microsomal membranes, nor does SRP have any effect on polypeptide chain elongation. All of these data are consistent with a posttranslational insertion mechanism for cytochrome b_5 and NADH cytochrome b_5 reductase. A model for the orientation of cytochrome b_5 (66) in the ER is illustrated in Figure 7.4.

Concluding Remarks

The above discussion has attempted to bring the reader up to date on specific ER membrane proteins that are inserted into the membrane either cotranslationally or posttranslationally. The absolute requirement for the SRP has been documented for cotranslational insertion of protein into the ER, confirming the original hypothesis of Blobel and coworkers (2, 5). However, it is still unclear why these proteins remain in the ER membrane, rather than being translocated across the membrane like secretory proteins or other organelle proteins. Although the lack of a cleavable NH_2 terminal sequence appeared to be a general characteristic of ER membrane proteins, the recent finding by MacKenzie and Owens (46) may exclude this possibility. Whether UDP-glucuronosyltransferase lacks a cleavable signal sequence remains to be confirmed by sequencing the primary translation product and the mature enzyme. Nevertheless, there does appear to be either structural constraints based on the conformation of the proteins or the presence of an additional sequence serving as a signal to terminate

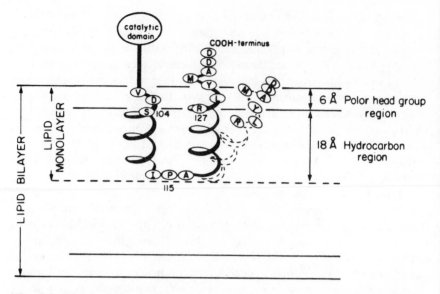

Fig. 7.4. Model of the tertiary structure of the membranous segment of rabbit cytochrome b_5 in the ER membrane. The tilting of the α-helix from Ala-116 to Met-130 represents an alternative orientation of cytochrome b_5 in the ER membrane. (This figure has been reprinted from [66] with permission of the authors.)

translocation that prevents the discharge of these proteins into the lumen of the ER. Future experiments will undoubtedly rely on the use of recombinant DNA techniques to construct hybrid proteins containing various regions of ER membrane proteins. These hybrid proteins can be utilized to examine the role that various domains of ER membrane proteins play in membrane insertion.

The assembly of rabbit liver cytochrome b_5 and steer liver cytochrome b_5 reductase into the ER membrane appears to be dependent upon their hydrophobic carboxyl and NH_2 terminal domains, respectively; thus, it would appear these sequences function as insertion sequences. The assembly of these proteins into the ER membrane is most likely governed by the thermodynamics of protein folding, which is a hypothesis proposed by Wickner and coworkers (1). The lack of membrane specificity for insertion of these proteins supports this hypothesis.

Acknowledgments. The authors would like to thank Joan Kiliyanski for her help in the preparation of this review.

References

1. Wickner, W. (1979) Ann. Rev. Biochem. *48*, 23–46.
2. Blobel, G. (1980) Proc. Natl. Acad. Sci. USA *77*, 1496–1500.
3. Kreil, G. (1981) Ann. Rev. Biochem. *50*, 317–348.

4. Sabatini, D.D., Kreibich, G., Morimoto, T. & Adesnik, M. (1981) J. Cell. Biol. *92*, 1–22.
5. Walter, P., Gilmore, R. & Blobel, G. (1984) Cell *38*, 5–8.
6. Omura, T., Siekevitz, P. & Palade, G. (1967) J. Biol. Chem. *242*, 2389–2396.
7. Arias, I.M., Doyle, D. & Schimke, R.T. (1969) J. Biol. Chem. *244*, 3303–3315.
8. Swick, R.W. & Ip, M.M. (1974) J. Biol. Chem. *249*, 6836–6841.
9. Parkinson, A., Thomas, P.E., Ryan, D.E. & Levin, W. (1983) Arch. Biochem. Biophys. *225*, 216–236.
10. Ryan, D.E., Thomas, P.E., Korzeniowski, D. & Levin, M. (1979) J. Biol. Chem. *254*, 1365–1374.
11. Ryan, D.E., Thomas, P.E. & Levin, W. (1982) Arch. Biochem. Biophys. *216*, 272–288.
12. Sadano, H. & Omura, T. (1983) J. Biochem. *93*, 1375–1383.
13. Shiraki, H. & Guengerich, F.P. (1984) Arch. Biochem. Biophys. *235*, 86–96.
14. Edwards, P.A. & Gould, R.G. (1972) J. Biol. Chem. *247*, 1520–1524.
15. Edwards, P.A., Lau, S.-F., Tanaka, R.D. & Fogelman, A.M. (1983) J. Biol. Chem. *258*, 7272–7275.
16. Negishi, M., Fujii-Kuriyama, Y., Tashiro, Y. & Imai, Y. (1976) Biochem. Biophys. Res. Commun. *71*, 1153–1160.
17. Takagi, M. (1977) J. Biochem. *82*, 1077–1084.
18. Ohlsson, R. & Jergil, B. (1977) Eur. J. Biochem. *72*, 595–603.
19. Dubois, R.N. & Waterman, M.R. (1979) Biochem. Biophys. Res. Commun. *90*, 150–157.
20. Bar-Nun, S., Kreibich, G., Adesnik, M., Alterman, L., Negishi, M. & Sabatini, D.D. (1980) Proc. Natl. Acad. Sci. USA *77*, 965–969.
21. Walz, F.G., Jr., Vlasuk, G.P., Omiecinski, L.J., Bresnick, E., Thomas, P.E., Ryan, D.E. & Levin, W. (1982) J. Biol. Chem. *257*, 4023–4026.
22. Pickett, C.B., Jeter, R.L., Wang, R. & Lu, A.Y.H. (1983) Arch. Biochem. Biophys. *225*, 854–860.
23. Fugii-Kuriyama, Y., Mizukami, Y., Kawajiri, K., Sagawa, K. & Muramatsu, M. (1982) Proc. Natl. Acad. Sci. USA *79*, 2793–2797.
24. Yuan, P.-M., Ryan, D.E., Levin, W. & Shively, J.E. (1983) Proc. Natl. Acad. Sci. USA *80*, 1169–1173.
25. Kumar, A. & Padmanaban, G. (1980) J. Biol. Chem. *255*, 522–525.
26. Bresnick, E., Brosseau, M., Levin, W., Reik, L., Ryan, D.E. & Thomas, P.E. (1981) Proc. Natl. Acad. Sci. USA *78*, 4083–4087.
27. Pickett, C.B., Telakowski-Hopkins, C.A., Donohue, A.M., Lu, A.Y.H. & Hales, B.F. (1982) Biochem. Biophys. Res. Commun. *104*, 611–619.
28. Morville, A.L., Thomas, P., Levin, W., Reik, L., Ryan, D.E., Raphael, C. & Adesnik, M. (1983) J. Biol. Chem. *258*, 3901–3906.
29. Kawajiri, K., Gato, O., Sogawa, K., Tagashira, Y., Muramatsu, M. & Fugii-Kuriyama, Y. (1984) Proc. Natl. Acad. Sci. USA *81*, 1649–1653.
30. Sogawa, K., Gotoh, O., Kawajiri, K. & Fugii-Kuriyama, Y. (1984) Proc. Natl. Acad. Sci. USA *81*, 5066–5070.
31. Gonzalez, F.J. & Kasper, C.B. (1980) Biochem. Biophys. Res. Commun. *93*, 1254–1258.
32. Pickett, C.B., Rosenstein, N.R., Jeter, R.L., Morin, J., & Lu, A.Y.H. (1980) Biochem. Biophys. Res. Commun. *94*, 542–548.
33. Pickett, C.B. & Lu, A.Y.H. (1981) Proc. Natl. Acad. Sci. USA *78*, 893–897.

34. Okada, Y., Frey, A.B., Guenthner, T.M., Oesch, F., Sabatini, D.D. & Kreibich, G. (1982) Eur. J. Biochem. *122*, 393–402.
35. Black, S.D. & Coon, M.J. (1982) J. Biol. Chem. *259*, 5938–5959.
36. Porter, T.D. & Kasper, C.B. (1985) Proc. Natl. Acad. Sci. USA *82*, 973–977.
37. Sakaguchi, M., Mihara, K. & Sato, R. (1984) Proc. Natl. Acad. Sci. USA *81*, 3361–3364.
38. Anderson, D.J., Walter, P. & Blobel, G. (1982) J. Cell Biol. *93*, 501–506.
39. Erickson, A.H., Walter, P. & Blobel, G. (1983) Biochem. Biophys. Res. Commun. *115*, 275–380.
40. Anderson, D.J., Mostov, K.E. & Blobel, G. (1983) Proc. Natl. Acad. Sci. USA *80*, 7249–7253.
41. Brown, D.A. & Simoni, R.D. (1984) Proc. Natl. Acad. Sci. USA *75*, 1674–1678.
42. Chin, D.J., Gil, G., Russell, D.W., Liscum, L., Luskey, K.L., Basu, S.K., Okayama, H., Berg, P., Goldstein, J.L. & Brown, M.S. (1984) Nature (London) *308*, 613–617.
43. Armstrong, R.N., Pinto-Coelho, C., Ryan, P.E., Thomas, P.E. & Levin, W. (1983) J. Biol. Chem. *258*, 2106–2108.
44. Liscum, L., Cummings, R.D., Anderson, R.G.W., DeMartino, G.N., Goldstein, J.L. & Brown, M.S. (1983) Proc. Natl. Acad. Sci. USA *80*, 7165–7169.
45. Liscum, L., Finer-Moore, J., Stroud, R.M., Luskey, K.L., Brown, M.S. & Goldstein, J.L. (1985) J. Biol. Chem. *260*, 522–530.
46. Mackenzie, P.I. & Owens, I.S. (1984) Biochem. Biophys. Res. Commun. *122*, 1441–1449.
47. Ito, A. & Sato, R. (1968) J. Biol. Chem. *243*, 4922–4923.
48. Spatz, L. & Strittmatter, P. (1971) Proc. Natl. Acad. Sci. USA *68*, 1042–1046.
49. Ozols, J. (1974) Biochemistry *13*, 426–434.
50. Enoch, H.G., Fleming, P.J. & Strittmatter, P. (1979) J. Biol. Chem. *254*, 6483–6488.
51. Remacle, J. (1978) J. Cell Biol. *79*, 291–313.
52. Spatz, L. & Strittmatter, P. (1973) J. Biol. Chem. *248*, 793–799.
53. Mihara, K. & Sato, R. (1975) J. Biochem. (Tokyo) *78*, 1057–1073.
54. Rogers, M.J. & Strittmatter, P. (1974) J. Biol. Chem. *249*, 5565–5569.
55. Mihara, K. & Sato, R. (1972) J. Biochem. (Tokyo) *78*, 1057–1073.
56. Okuda, T., Mihara, K. & Sato, R. (1972) J. Biochem. (Tokyo) *72*, 987–992.
57. Mihara, R., Sato, R., Sakakibara, R. & Wada, H. (1978) Biochemistry *17*, 2829–2834.
58. Kensil, C.R., Hediger, M.A., Ozols, J. & Strittmatter, P. (1983) J. Biol. Chem. *258*, 14656–14663.
59. Ozols, J., Carr, S.A. & Strittmatter, P. (1984) J. Biol. Chem. *259*, 13349–13354.
60. Harano, T. & Omura, T. (1977) J. Biochem. (Tokyo) *82*, 1541–1549.
61. Elhammer, A.G., Dallner, G. & Omura, T. (1978) Biochem. Biophys. Res. Commun. *84*, 572–580.
62. Borgese, N. & Gaetani, S. (1980) FEBS Lett. *112*, 216–220.
63. Rachubinski, R.A., Verma, D.P.S. & Bergeron, J.J.M. (1980) J. Cell Biol. *84*, 705–716.
64. Borgese, N. & Gaetani, S. (1983) EMBO J. *2*, 1263–1269.
65. Bendzko, P., Puehn, S., Pfeil, W. & Rapoport, T.A. (1982) Eur. J. Biochem. *123*, 121–126.
66. Takagaki, Y., Radhakrishnan, R., Gupta, C. & Khorana, H.G. (1983) J. Biol. Chem. *259*, 9128–9135.

67. Haniu, M., Ryan, D.E., Iida, S., Lieber, C.S., Levin, W. & Shively, J.E. (1984) Arch. Biochem. Biophys. *235*, 304–311.
68. DuBois, G.C., Apella, E., Armstrong, R., Levin, W., Lu, A.Y.H. & Jerina, D.M. (1979) J. Biol. Chem. *254*, 6240–6243.
69. Heinemann, F.S. & Ozols, J. (1984) J. Biol. Chem. *259*, 797–804.
70. Tarr, G.E., Black, S.D., Fujita, V.S. & Coon, M.J. (1983) Proc. Natl. Acad. Sci. USA *80*, 6552–6556.

8

Import of Proteins Into Mitochondria

Richard Zimmermann

Mitochondria are the only organelles in animal and fungal cells with their own genome and the machinery for its expression. Despite this fact, the majority of mitochondrial proteins are coded by the nuclear genome and synthesized on cytoplasmic ribosomes (1). These proteins must be delivered to their functional location in outer membrane, inner membrane, intermembrane space, or matrix (Figure 8.1). Therefore, central problems in mitochondrial biogenesis are: 1) How are nuclear-coded mitochondrial proteins guided to their correct location in the mitochondrion? 2) How is the insertion and/or transfer of these proteins into or across the mitochondrial membrane(s) accomplished? 3) How do imported proteins acquire their final functional characteristics?

Posttranslational Import of Proteins Into Mitochondria

Two distinct mechanisms for conducting proteins through or into membranes appear to operate in the eukaryotic cell (Figure 8.2). In one mechanism, membrane choice and membrane insertion and/or transfer are coupled to translation (cotranslational mechanism). This type of pathway seems to be followed by secreted proteins, and at least by some proteins of plasma membrane, Golgi apparatus, lysosomes, and endoplasmic reticulum (ER). The synthesis of these proteins is initiated on free ribosomes. As elongation continues, membrane choice occurs and the translating polysomes become bound to the membrane of the ER. The translation products are inserted into or transferred across this membrane during elongation. Afterwards, the proteins are delivered from the ER to their functional location concomitant with the flow of membranes. Alternatively, precursor proteins may be synthesized on free polysomes and transported to the target membrane or organelle after synthesis is completed (posttranslational

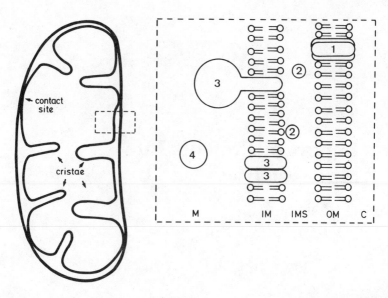

Fig. 8.1. Schematic diagram of a mitochondrion. 1. Outer membrane protein, example: porin. 2. Intermembrane space protein, examples: cytochrome B_2 and cytochrome C. 3. Inner membrane protein, examples: F_0/F_1-ATPase and ADT/ATP carrier. 4. Matrix protein, example: ornithine transcarbamylase. OM, outer membrane; IMS, intermembrane space; IM, inner membrane; M. matrix; C, cytosol.

mechanism). This type of pathway is followed by proteins of mitochondria and chloroplasts, and at least by some proteins of peroxisomes and glyoxysomes. In this case, precursor proteins move from the cytosol to their target membrane or organelle after completion of their synthesis. Proteins of mitochondria and chloroplasts are then specifically distributed between membranous or aqueous compartments within these organelles.

The first evidence for a posttranslational mechanism in transport of proteins into mitochondria was obtained by pulse chase experiments with intact cells of *Neurospora crassa* and *Saccaromyces cerevisiae* (2, 3). The striking observations in these studies were: 1) there was a lag in the appearance of labeled proteins in the mitochondria compared to the cytosol; 2) there were different kinetics for the appearance of different proteins within the mitochondria; 3) inhibition of protein synthesis did not block import of proteins into mitochondria; and 4) pools of mitochondrial proteins were detected outside the mitochondria. Since the emergence of these purely descriptive data, we have learned many details about the system and have gained some insights into the mechanisms involved (4, 5). This is mainly due to the impact of *in vitro* systems for protein synthesis and transport (3, 6). However, most insights obtained *in vitro* have been substantiated in experiments with intact cells.

The general scheme for import of nuclear-coded proteins into mitochondria seems to be that precursors of these proteins are synthesized on free cytoplasmic

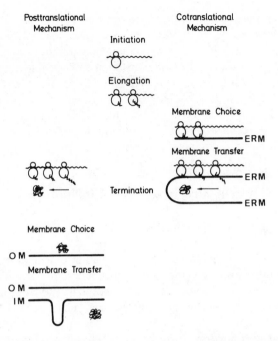

Fig. 8.2. Mechanisms for the transfer of proteins across membranes. OM, mitochondrial outer membrane; IM, mitochondrial inner membrane; and ERM, membrane of the endoplasmic reticulum.

ribosomes and released into the cytosol. The first step in import is binding of the precursors to the mitochondria. This binding is followed by membrane insertion and/or transfer. For many proteins, this second step depends on the membrane potential across the inner mitochondrial membrane, and is followed by proteolytic processing of the imported precursor by a matrix-localized protease.

In the following sections, the experimental findings that have produced this picture are discussed.

Site of Synthesis of Nuclear-Coded Mitochondrial Proteins

Considerable effort has been expended to determine the types of cytoplasmic ribosomes that synthesize mitochondrial proteins. For this purpose, cytoplasmic polysomes were usually fractionated into free polysomes and polysomes associated with microsomal or mitochondrial membranes (7). The polysomes or their mRNA contents were then incubated in *in vitro* systems to allow synthesis of the proteins that the polysomes were synthesizing at the time of their isolation. In every case, when specific mitochondrial proteins were analyzed, synthesis occurred predominantly on free cytoplasmic ribosomes (10, 16, 23, 30, 31, 49, 75, 77, 80, 90). Depending on the fractionation conditions, polysomes synthesizing mitochondrial proteins have been found in association with mitochondrial and microsomal membranes to varying extents (reviewed in [5]). However, under no circumstances could synthesis on membrane-bound ribosomes be characterized as a prerequisite for transport of the translation products into mitochondria. The lesson from these studies seems to be that synthesis of mitochondrial proteins can occur on free, as well as membrane-bound, ribosomes; but there is clearly no coupling between synthesis on membrane-bound ribosomes and insertion of the translation products into the relevant membrane (see also below). This is true for proteins of all mitochondrial compartments.

Extramitochondrial Precursors of Mitochondrial Proteins

Location of the Extramitochondrial Precursors

There are two lines of evidence that precursors of mitochondrial proteins are released into the cytosol. Precursors of mitochondrial proteins have been found in the cytosol after pulse labeling of intact cells and fractionation of the cells (44, 45, 73, 76, 80, 81, 84). Pulse chase analysis indicated that the kinetics of cytosolic precursors' appearance and disappearance were consistent with their being en route to the mitochondria. Furthermore, after synthesis of mitochondrial proteins *in vitro*, the translational products can be recovered in a postribosomal supernatant. When this equivalent of a cellular cytosol is incubated with mitochondria, the mitochondrial precursors are imported into mitochondria (see the section in this chapter on "Membrane Transfer in Import of Mitochondrial Proteins"). These observations support the conclusion that synthesis occurs on free ribosomes and not on membrane-bound ribosomes, and they are in agreement with the earlier data suggesting that proteins are imported into mitochondria posttranslationally. As with synthesis on free cytosplasmic ribosomes, release of newly synthesized precursors into the cytosol seems to be a general rule for mitochondrial proteins.

Nature of the Extramitochondrial Precursors

Currently available data on the characteristics of the cytosolic precursors of mitochondrial proteins are presently sparse and almost totally descriptive. The available observations from *in vivo* and *in vitro* studies are: 1) *Some precursors are larger in molecular weight than their mature counterparts and some are not* (Table 8.1). Strikingly, the proteins of the outer membrane are synthesized with the same molecular weight; the proteins of the inner membrane and the matrix are synthesized with a larger molecular weight. However, there are exceptions to these rules. The proteins of the intermembrane space seem to be heterogeneous in this respect. They are either not proteolytically processed at all or processed in two steps (27, 29). The apparent molecular weight of the additional sequences

Table 8.1. Precursors of Mitochondrial Proteins

Mitochondrial Compartment	Protein	Organism	Apparent precursor	MW (kd$_a$) Mature	Membrane potential-dependent import	References
OM	Monoamine oxidase	Rat	59	59		8
	OMM-35	Rat	35.5	35		9
	Porin	N.c.	31	31	−	10
	Porin	Yeast	29	29	−	11, 12
	Protein (14 kd)	Yeast	14	14	−	13
	Protein (45 kd)	Yeast	45	45	−	13
	Protein (70 kd)	Yeast	70	70	−	13
IMS	Adenylate Kinase	Chicken	28	28		14
	Cytochrome C	N.c.	12	12	−	15–20
	Cytochrome C	Horse	12	12		21
	Cytochrome C	Yeast	12	12		22
	Cytochrome C	Rat	12	12		23
	Cytochrome C peroxidase	Yeast	39.5	33.5		24–26
	Cytochrome B$_2$	Yeast	68	58	+	26–29
	Sulfite oxidase	Rat	59	55		30
IM	ADP/ATP carrier	N.c.	32	32	+	31–36
	F$_1$-ATPase α-subunit	Yeast	64	58		37–45, 54
	F$_1$-ATPase β-subunit	Yeast	56	54	+	37–45, 54
	F$_1$-ATPase γ-subunit	Yeast	40	34		37–45, 54
	F$_1$-ATPase β-subunit	N.c.	58	56	+	46
	F$_1$-ATPase δ-subunit	N.c.	20	15–16		111
	F$_0$-ATPase subunit 9	N.c.	12–16.4	8–10.5	+	47–51
	Carnitine acetyltransferase	Rat	69	67.5		52
	Cytochrome P-450$_{scc}$	Cattle	54.5	49		53
	Cytochrome C oxidase					
	Subunit IV	Yeast	17	14		54–58
	Subunit V	Yeast	15	12.5		54–58
	Subunit VI	Yeast	17–20	12.5		54–58
	Subunit VII	Yeast	5–7.5	5–7.5		54–58
	Subunit IV	Rat	18–19.5	16.5		59, 60
	Subunit V	Rat	15.5	12.5		59, 60
	D-β-Hydroxybutyrate-dehydrogenase	Rat	37	32		30
	Ubiquinol:cytochrome C reductase					
	Subunit I	Yeast	44.5	44		61–67
	Subunit II	Yeast	40.5	40		61–67
	Subunit V	Yeast	27	25		61–67
	Subunit VI	Yeast	26	17		61–67
	Subunit VII	Yeast	14	14		61–67
	Subunit VIII	Yeast	11	11		61–67
	Cytochrome C$_1$	Yeast	37	31	+	61–67
	Subunit I	N.c.	51.5	50	+	68
	Subunit II	N.c.	47.5	45	+	68
	Subunit V	N.c.	28	25	+	68

Table 8.1. (Continued)

Mitochondrial Compartment	Protein	Organ-ism	Apparent pre-cursor	MW (kd$_a$) Mature	Membrane potential-dependent import	References
	Subunit VI	N.c.	14	14		68
	Subunit VII	N.c.	12	11.5	+	68
	Subunit VIII	N.c.	11.6	11.2		68
	Cytochrome C$_1$	N.c.	38	31	+	68
	Uncoupling protein	Rat	32	32		69
M	Adrenodoxin	Cattle	20	12		70
	Adrenodoxin reductase	Cattle	50	50		70
	δ-Aminolevulinate synthase	Rat	51	45		71–73
	δ-Aminolevulinate synthase	Chicken	75	63–65		71–73
	Aspartate amino-transferase	Rat	47	45		74–76, 88
	Aspartate amino-transferase	Chicken	47	44.5	+	74–76, 88
	Carbamoyl-phosphate synthetase	Rat	165a	160		77–84
	Carbamoyl-phosphate synthetase	Frog	160	160		78
	Citrate synthase	Yeast	52	50		57
	Citrate synthase	N.c.	47	45		85
	L-Glutamate dehydrogenase	Rat	60b	54		30, 86, 87
	2-Isopropylmalate synthase	Yeast	65	65	+	89
	Malate dehydrogenase	Rat	38	37		30
	• Ornithine amino-transferase	Rat	49	43		90
	Ornithine trans-carbamylase	Rat	39.5–43	36–39	+	83, 84, 91–101
	RNA polymerase	Yeast	47	45		102
	Mn^{2+} superoxide dismutase	Yeast	26	24	+	103

Abbreviations: OM, outer membrane; IMS, intermembrane space; IM, inner membrane; M, matrix; and N.c., neurospora crassa.
[a] Enzymatically inactive.
[b] Enzymatically active.

varies widely, between 500 and 10,000. There clearly is no obvious correlation between the size of an additional sequence and a mature protein's location in the mitochondrion. In all cases where it has been directly looked at, the additional sequence was found to be at the amino terminus. Thus far, four additional sequences have been determined (Table 8.2). 2) *Many precursor proteins have basic isoelectric points* (104) In many cases, this is due to the amino acid composition of the additional sequence (Table 8.2) (104). 3) *At least some precursor proteins are present in the cytosol in the form of protein aggregates.*

The general rule behind these features may be that a precursor protein has to be kept in a conformation differing from that of the mature protein. For an individual precursor protein, this may be accomplished by a different size and/or a difference in the state of aggregation. The other general rule may be that the precursors have to be positive in net charge or have a basic domain.

Table 8.2. Aminoterminal Sequences of Precursors of Mitochondrial Proteins

Protein/ Organism	Aminoterminal sequence of precursor	References
F$_0$-ATPase subunit 9 Amino acids 1–75 Neurospora crassa	Met-Ala-Ser-Thr-Arg-Val-Leu-Ala-Ser-Arg-Leu-Ala-Ser-Gln-Met- -Ala-Ala-Ser-Ala-Lys-Val-Ala-Arg-Pro-Ala-Val-Arg-Val-Ala-Gln- -Val-Ser-Lys-Arg-Thr-Ile-Gln-Thr-Gly-Ser-Pro-Leu-Gln-Thr-Leu- -Lys-Arg-Thr-Gln-Met-Thr-Ser-Ile-Val-Asn-Ala-Thr-Thr-Arg-Gln- -Ala-Phe-Gln-Lys-Arg-*Ala-Tyr*-Ser-Ser-Glu-Ile-Ala-Gln-Ala-Met-	48
F$_1$-ATPase δ-subunit Amino acids 1–45 Neurospora crassa	Met-Asn-Ser-Leu-Arg-Ile-Ala-Arg-Ala-Ala-Ala-Leu-Arg-Val-Arg- -Pro-Thr-Ala-Val-Arg-Ala-Pro-Leu-Gln-Arg-Arg-<u>Gly-Tyr</u>-Ala- -Glu-Ala-Val-Ala-Asp-Lys-Ile-Lys-Leu-Ser-Leu-Ser-Leu-Pro-His-	111
Cytochrome C-peroxidase Amino acids 1–75 Yeast	Met-Thr-Thr-Ala-Val-Arg-Leu-Leu-Pro-Ser-Leu-Gly-Arg-Thr-Ala- -His-Lys-Arg-<u>Ser-Leu</u>-Tyr-Leu-Phe-Ser-Ala-Ala-Ala-Ala-Ala-Ala- -Ala-Ala-Ala-Ala-Thr-Phe-Ala-Tyr-Ser-Gln-Ser-His-Lys-Arg-Ser- -Ser-Ser-Ser-Pro-Gly-Gly-Gly-Ser-Asn-His-Gly-Trp-Asn-Asn-Trp- -Gly-Lys-Ala-Ala-Ala-Leu-Ala-**Ser-Thr**-Thr-Pro-Leu-Val-His-Val-	25
Ubiquinol: Cytochrome C Reductase subunit VI Complete sequence Yeast	Met-Asp-Met-Leu-Glu-Leu-Val-Gly-Glu-Tyr-Trp-Glu-Gln-Leu-Lys- -Ile-Thr-Val-Val-Pro-Val-Val-Ala-Ala-Ala-Glu-Asp-Asp-Asp-Asn- -Glu-Gln-His-Glu-Glu-Lys-Ala-Ala-Glu-Gly-Glu-Glu-Lys-Asp-Asp- -Asp-Asn-Gly-Asp-Glu-Asp-Glu-Asp-Glu-Asp-Glu-Asp-Glu-Asp-Asp- -Asp-Asp-Asp-Asp-Asp-Glu-Asp-Glu-Glu-Glu-Glu-Glu-Glu-Val-Thr- -Asp-Gln-Leu-Glu-Asp-Leu-Arg-Glu-His-Phe-Lys-Asn-Thr-Glu-Glu- -Gly-Lys-Ala-Leu-Val-His-His-Tyr-Glu-Glu-Cys-Ala-Glu-Arg-Val- -Lys-Ile-Gln-Gln-Gln-Gln-Pro-Gly-Tyr-Ala-Asp-Leu-Glu-His-Lys- -Glu-Asp-Cys-Val-Glu-Glu-Phe-Phe-His-Leu-Gln-His-Tyr-Leu-Asp- -Thr-Ala-Thr-Ala-Pro-Arg-Leu-Phe-Asp-Lys-Leu-Lys	64

Italic letters indicate cleavage site for protease 1, matrix enzyme, phenantroline-sensitive.
Boldface letters indicate cleavage site for protease 2, intermembrane space or inner membrane enzyme, phenantroline-resistant.
Underscored letters indicate cleavage site for protease 1, predicted.
For further sequences, refer to references 113–123.

Organelle Choice in Import of Mitochondrial Proteins

Choice of the ER membrane is achieved by the cytoplasmic signal recognition particle (SRP) and its receptor on the ER—the docking protein—for proteins following the cotranslational pathway (Figure 8.2), as became clear through the elegant work in the laboratories of Blobel and Dobberstein. The fundamental difference between this mechanism and membrane choice in the posttranslational pathway (Figure 8.2) is that in the latter, complete precursor proteins have to choose or be chosen by the correct membrane. Therefore, a different mechanism would not be unexpected in this case.

Receptors on the Mitochondrial Surface

It was possible for a time to demonstrate posttranslational import of proteins into mitochondria *in vitro* before binding of the precursors to the mitochondria was appreciated as being the first step in this import. This realization stemmed from the development of conditions where binding could be separated from import through the action of certain inhibitors.

There is very good evidence that there is a proteinaceous receptor for apocytochrome C on the mitochondrial surface, which binds apocytochrome C prior to its translocation across the outer membrane and conversion to holocytochrome C (15–23). When apocytochrome C is synthesized *in vitro* and incubated with mitochondria in the presence of the heme analog, deuterohemin apocytochrome C becomes bound to the outer surface of the mitochondria (18). This binding shows the accepted characteristics for a specific receptor ligand interaction: 1) *The binding is rapid and reversible.* Bound apocytochrome C can be displaced by excess of mitochondrial apocytochrome C from various species, but not by holocytochrome C or apocytochrome C from the bacterium *Paracoccus denitrificans* (19). 2) *The binding sites are present in limited number (90 pmol/mg of mitochondrial protein)* (19). 3) *Receptor-bound apocytochrome C is imported into mitochondria and converted to holocytochrome C when inhibition by deuterohemin is relieved by excess of hemin* (18). The binding site is a protein, since it can be eliminated by treating the mitochondria with protease prior to incubation in the *in vitro* system (20). Apocytochrome C is not synthesized with an additional amino terminal sequence (16), and it can be isolated in chemical amounts from holocytochrome C (15). With this isolated precursor protein, it was shown that the apocytochrome C receptor is not used by the precursors of ADP/ATP carrier (17), cytochrome c_1 (68), and subunit 9 of the ATPase (17). This indicates that different receptors exist. It was also shown with this protein that the domain of apocytochrome C recognizing the receptor is within the carboxy terminal half of the protein (23). The availability of chemical amounts of the precursor are currently used to isolate the receptor protein from *N. crassa* (H. Koehler, B. Henning, and W. Neupert, personal communication).

Experiments with inhibitors of the oxidative phosphorylation have provided evidence for at least two more proteins that binding to a receptor is involved in import. These proteins are the ADP/ATP carrier in *Neurospora crassa* (33) and cytochrome B_2 in yeast (28). In both cases, a specific binding to mitochondria was observed in the presence of uncouplers, and the binding was found to be sensitive to pretreatment of the mitochondria with protease (28, 33, 36). For cytochrome B_2, the specific binding could also be observed with outer membrane vesicles or liposomes made with outer membrane extract (28). There are other proteins that a receptor involvement has been concluded for from experiments with proteased mitochondria. However, it is not clear whether indirect effects of this protease treatment have been ruled out in every case. This criterion seems to be reasonable only when specific binding can be demonstrated for untreated mitochondria.

The situation for outer membrane proteins is unclear. There is evidence for a protease-sensitive site involved in porin insertion into *Neurospora* mitochondria (36) but this is not the case for the yeast porin (12).

Soluble Cytoplasmic Factors

There is evidence for the involvement of cytoplasmic factors, at least for two proteins: ornithine transcarbamylase and β-subunit of ATPase (100, 101, 112). However, it is still unclear whether the effect is on membrane choice or if it has to do with another step in import. It could be that the observed cytoplasmic factors are involved in keeping the precursors in solution or in the right conformation, as was speculated for the aggregation or synthesis as larger precursors. There is too little information available thus far to conclude how general such a requirement is.

Membrane Transfer in Import of Mitochondrial Proteins

For many mitochondrial proteins, the membrane potential across the inner membrane is required for import (Table 8.1) (35, 38, 99, 104). There seem to be two types of import with respect to the potential. 1) *Cytochrome C peroxidase and cytochrome B_2 are synthesized as larger precursors and are processed by the protease in the matrix* (see the section in this chapter on "Covalent Modifications") *and as a second step by a protease in the intermembrane space* (27, 29). *These proteins require a membrane potential for import. The same seems to be true for all proteins of the inner membrane and the matrix, including the few proteins of these locations that are not synthesized as larger precursors.* 2) *Proteins of the outer membrane and some proteins of the intermembrane space do not require an energized mitochondrial inner membrane for import.* In the case of the intermembrane space proteins, this is true for cytochrome C. This protein is not synthesized as a larger precursor, as is true for the proteins of the outer membrane

(Table 8.1). The lesson from these studies seems to be that an energized mitochondrial membrane is required for all proteins that partly or completely cross the inner membrane on the way to their functional location. This includes certain proteins of the intermembrane space, which for some unknown reason are processed by the matrix protease (27, 29) and therefore have to become exposed to the inside of the inner membrane, at least with the domain to be processed. The membrane potential obviously is not essential for the insertion of proteins into or across the outer membrane. This indicates that there are at least two pathways for proteins to be transported into mitochondria.

There is an additional piece of information available on the transfer mechanism. This is that the precursors of β-subunit of ATPase in *Neurospora crassa* (46) and cytochrome C peroxidase in yeast (26), both with additional sequences, are imported into mitochondria even in the absence of processing. When processing is inhibited or slowed, unprocessed precursor proteins are found within the organelle (26, 46). From this, one can conclude that transport is not obligately coupled to processing. This makes it unlikely, at least for these proteins, that membrane choice and transfer are driven by the covalent modification. On the other hand, apocytochrome C is not imported into mitochondria, but stays bound to its receptor when the heme attachment is inhibited by the heme analog deuterohemin (18). In this case, processing is coupled with import (see the section below on "Covalent Modifications").

Covalent Modifications of the Imported Mitochondrial Proteins and Assembly with Functional Complexes

Covalent Modifications

Two types of covalent modifications have been observed being related to import of proteins into mitochondria. 1) *There are many proteins that are synthesized as larger precursors* (Table 8.1). *These proteins are processed to their mature molecular weight by mitochondrial proteases specifically removing the additional sequence in one or two steps* (see below). 2) *Some proteins get modified by covalent attachment of a prosthetic group, such as cytochromes C and C_1.* In the case of cytochrome C, the heme attachment is the only modification (16), in the case of cytochrome C_1, there are additional modifications by proteases (29).

The problem of covalent modifications of proteins during or after transport into mitochondria first has to be seen in connection with the nature of precursors. If it is a prerequisite for import to keep precursor proteins in a conformation different from the conformation of the mature protein, there have to be mechanisms to trigger the conformational change, during or after import. The actual trigger of the conformational change may be either the modifying reaction itself or a preceding step in import, such as interaction with a receptor or the membrane itself. Second, covalent modifications and/or conformational changes may have to do with the assembly of the imported proteins to structures of a higher order.

For a particular protein, both themes might exist in parallel. The general rule may be that for some proteins, it is necessary to make the conformational change irreversible by way of a covalent modification. How this is linked to the actual transfer mechanism can only be speculated. It is possible that for some proteins, the conformational change makes the import unidirectional or irreversible and is therefore linked to import. For other proteins, the import could be irreversible anyway; *e.g.*, because of the membrane potential. In this case, one would expect modification and import to be independent of each other. The modification would be necessary only to make the conformational change irreversible.

According to these considerations for apocytochrome C, which is imported into mitochondria independent of the membrane potential, the covalent attachment of heme is necessary for the conformational change to be irreversible. Furthermore, the conformational change is necessary for import to be irreversible. Therefore, import of apocytochrome C is coupled to the conversion to holocytochrome C. This is not the case for cytochrome C_1 (29). For proteins such as cytochrome C peroxidase and β-subunit of the ATPase, which are synthesized with an additional sequence and import-dependent on the membrane potential, the processing is not coupled to import into mitochondria (26, 46). Conformational change and/or modification in these cases seem to be independent of import. However, for cytochrome C peroxidase, two processing steps seem to be a prerequisite for the proper localization of the protein into the intermembrane space (29). For porin, which is imported into the outer membrane without proteolytic processing and without the aid of the membrane potential, there seems to be a conformational change concomitant with insertion into the outer membrane (12). For the ADP/ATP carrier, which is imported into the inner membrane without proteolytic processing but with the aid of the membrane potential, there seems to be a conformational change concomitant with insertion into the inner membrane (34). It is unclear whether in the last two cases conformational changes render the insertion irreversible or whether in the case of the ADP/ATP carrier, the potential is doing so.

Processing Enzymes

There is good evidence that there is one protease in the matrix (protease 1). It has been partially purified from yeast (41, 42, 57, 58) *Neurospora crassa* and rat liver (92, 96–98). In all cases, the protease was found to be a soluble enzyme in the matrix and to be sensitive to chelating agents. It is not clear whether this is the only protease that processes precursors of inner membrane and matrix, as well as cytochrome B_2 and cytochrome C peroxidase.

There is indirect evidence that there is a second protease (protease 2) in the intermembrane space or the inner membrane that cleaves the intermediates of cytochrome B_2, cytochrome C peroxidase, and cytochrome C_1. This enzyme may be soluble in the intermembrane space or on the outside of the inner membrane. It is resistant to chelating agents (27).

There is indirect evidence that the linkage of heme to cytochrome C is mediated by an enzyme, termed cytochrome C heme lyase, and that this enzyme is located in the intermembrane space (26).

Assembly

Conformation of the mature polypeptide, in most cases, means conformation prior to the assembly of the polypeptide into higher structures of either identical or different subunits. Some mitochondrial proteins contain polypeptides that are synthesized on mitochondrial ribosomes. How far some of the observed covalent modifications are related to assembly of higher structures is unclear at present. There is evidence, in at least some cases, that polypeptides imported into mitochondria *in vitro* are assembled into functional protein complexes (12, 34, 49, 82, 94).

Working Models for the Import of Proteins Into Mitochondria

Figure 8.3 summarizes in pictorial form our current knowledge of the process of protein import into mitochondria. Nuclear-coded mitochondrial proteins are synthesized on free cytoplasmic ribosomes and released into the cytosol.

    1. Outer membrane proteins are inserted into the outer membrane only. They are a homogeneous class of proteins (Table 8.1) in that they are synthesized

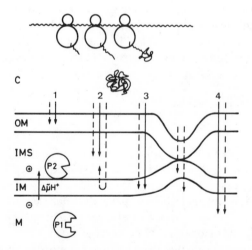

Fig. 8.3. Pathways for the import of proteins into mitochondria. 1 = outer membrane protein; 2 = intermembrane space protein; 3 = inner membrane protein; 4 = matrix protein. C, cytosol; OM, outer membrane; IMS, intermembrane space; IM, inner membrane; M, matrix; P1, protease 1 (see Table 8.2); P2, protease 2 (see Table 8.2); →, theoretical pathway; →, proposed pathway; and $\triangle\bar{\mu}H^+$, membrane potential.

without an additional sequence and are imported independent of the membrane potential across the inner membrane. It is unclear whether there are receptor proteins or other proteinaceous factors on the mitochondrial surface or from the cytosol involved in the insertion. The insertion might simply be partitioning into the lipid bilayer. In this case, the specificity could be due to the lipid composition of the outer membrane or to interaction with preexisting proteins.

   2. Theoretically, intermembrane space proteins only need to be transferred across the outer membrane. However, it appears that they reach the intermembrane space following one of two possible pathways. They are either transferred only across the outer membrane (cytochrome C type) or first associate with the inner membrane in the first steps of import, and are eventually converted into soluble proteins in the intermembrane space by a second processing step (cytochrome B_2 type). Cytochrome C-type proteins are similar to outer membrane proteins, in that they are synthesized without an additional sequence and are imported independent of the presence of a potential across the inner membrane (Table 8.1). Adenylate kinase seems to be another member of this class. In the case of cytochrome C, this process is mediated by a receptor protein on the mitochondrial surface. Cytochrome B_2-type proteins are similar to the majority of inner membrane proteins, in that they are synthesized as larger precursors, imported in a membrane potential-dependent manner, and processed by the matrix-processing enzyme (protease 1) (Table 8.1). A second processing in the intermembrane space (protease 2) leads to release of the proteins into the inner membrane space. A receptor protein appears to be involved in import of cytochrome B_2-type proteins. Another protein following this pathway is cytochrome C peroxidase.

   3. Inner membrane proteins have to be transferred across the outer membrane and inserted into the inner membrane. The simplest way of doing this would be on a contact site of outer and inner membrane. This would reduce the problems of choice of mitochondria and additional choice of a subcompartment for a given precursor. Inner membrane proteins, as well as the cytochrome B_2-type proteins of the intermembrane space, are usually synthesized as larger precursors, and they show a dependency on the membrane potential for import. They are usually processed to their mature molecular weight by a matrix protease (protease 1). There seems to be receptor involvement in import for some of these proteins, but this cannot be generalized at present. The same is true regarding the involvement of soluble cytoplasmic factors. All the proteins of the inner membrane studied thus far seem to be in agreement with the above statements except for some proteins such as the ADP/ATP carrier, which are not synthesized as larger precursors.

   4. Matrix proteins have to be transferred across both the inner and outer membranes. As for inner membrane proteins, this can be visualized as a one-step mechanism at contact sites. All features mentioned for inner membrane proteins are equally true for the matrix proteins.

   From this summary, it becomes clear how little we understand the actual mechanism of transport of proteins into mitochondria. There are distinct pathways becoming apparent, and some proteins involved in the pathways have been defined.

Open Questions

The following questions concerning the transfer of proteins across or through mitochondrial membranes need to be answered in the future:

1. How many different pathways are there? Even though we know many details about precursor-proteins of mitochondrial proteins and their conversion to the mature forms, there is more to be learned:

   a. What is the role of the additional sequences? Is the interaction a protein-lipid interaction for outer membrane proteins, but a protein-protein interaction for all other proteins? How many different receptors are there? What is the role of the soluble factors? Are they truly necessary for import or are they only necessary in the *in vitro* systems (124–1263)?

   b. How are the precursors with additional sequences processed by the specific proteases? What makes the cleavage so specific? What is the role of the processing?

   c. How do precursors interact with the mitochondria (127)?

2. Do contact sites between inner and outer membranes have a role in import? In other words, is there a one-step or a multiple-step event in import into the inner membrane and the matrix? Are proteins or the membrane potential involved in forming these contact sites (12632)?

3. What, if any, mitochondrial machinery exists to mediate the transfer of proteins across or into membranes (128, 129)? Are there pore-forming proteins involved or proteins facilitating the transfer in other ways? How is the membrane phospholipid involved in this transfer (as suggested for apocytochrome C import) (105)?

4. What is the role of the membrane potential? Is it acting directly by providing the energy for transfer (as proposed for the potential effect on bacterial secretion) (105–110)? Or, is it acting indirectly on proteins involved in the pathway or the membrane state in general (130)?

5. Last, but not least, there are many questions to be asked about assembly and regulation of import and assembly.

References

1. Schatz, G. & Mason, T.L. (1974) Ann. Rev. Biochem. *43*, 51–87.
2. Hallermayer, G., Zimmermann, R. & Neupert, W. (1977) Eur. J. Biochem. *81*, 523–532.
3. Schatz, G. (1979) FEBS Lett. *103*, 201–211.
4. Neupert, W. & Schatz, G. (1981) Trends Biochem. Sci. *6*, 1–4.
5. Schatz, G. & Butow, R.A. (1983) Cell *32*, 316–318.

6. Harmey, M.A., Hallermayer, G., Korb, H. & Neupert, W. (1977) Eur. J. Biochem. *81*, 533-544.

7. Suissa, M. & Schatz, G. (1982) J. Biol. Chem. *257*, 13048-13055.

8. Sagara, Y. & Ito, A. (1982) Biochem. Biophys. Res. Commun. *109*, 1102-1107.

9. Shore, G.C., Power, F., Bendayan, M. & Carignan, P. (1981) J. Biol. Chem. *256*, 8761-8766.

10. Freitag, H., Janes, M. & Neupert, W. (1982) Eur. J. Biochem. *126*, 197-202.

11. Mihara, K., Blobel, G. & Sato, R. (1982) Proc. Natl. Acad. Sci. USA *79*, 7102-7106.

12. Gasser, S.M. & Schatz, G. (1983) J. Biol. Chem. *258*, 3427-3430.

13. Riezman, H., Hase, T., van Loon, A.P.G.M., Grivell, L.A., Suda, K. & Schatz, G. (1983) EMBO J. *2*, 2161-2168.

14. Watanabe, K. & Kubo, S. (1982) Eur. J. Biochem. *123*, 587-592.

15. Korb, H. & Neupert, W. (1978) Eur. J. Biochem. *91*, 609-620.

16. Zimmermann, R., Paluch, U. & Neupert, W. (1979) FEBS Lett. *108*, 141-146.

17. Zimmermann, R., Hennig, B. & Neupert, W. (1981) Eur. J. Biochem. *116*, 455-460.

18. Hennig, B. & Neupert, W. (1981) Eur. J. Biochem. *121*, 203-212.

19. Hennig, B., Koehler, H. & Neupert, W. (1983) Proc. Natl. Acad. Sci. USA *80*, 4963-4967.

20. Hennig, B., Koehler, H. & Neupert, W. (1983) In Mitochondria 1983: Nucleo-cytoplasmic Interactions, 551-561, Schweyen, R.J., Wolf, K. & Kaudewitz, F., eds. DeGruyter, Berlin, New York.

21. Basile, G., Dibello, C. & Taniuchi, H. (1980) J. Biol. Chem. *255*, 7181-7191.

22. Veloso, D., Basile, G. & Taniuchi, H. (1980) J. Biol. Chem. *256*, 8646-8651.

23. Matsuura, S., Arpin, M., Hannum, C., Margoliash, E., Sabatini, D.D. & Morimoto, T. (1981) Proc. Natl. Acad. Sci. USA *78*, 4368-4372.

24. Maccecchini, M.L., Rudin, Y. & Schatz, G. (1979) J. Biol. Chem. *254*, 7468-7471.

25. Kaput, J., Goltz, S. & Blobel, G. (1982) J. Biol. Chem. *257*, 15054-15058.

26. Reid, G.A., Yonetani, T. & Schatz, G. (1982) J. Biol. Chem. *257*, 13068-13074.

27. Daum, G., Gasser, S.M. & Schatz, G. (1982) J. Biol. Chem. *257*, 13075-13080.

28. Riezman, H., Hay, R., Witte, Ch., Nelson, N. & Schatz, G. (1983) EMBO J. *2*, 1113-1118.

29. Gasser, S.M., Ohashi, A., Daum, G., Boehni, P.C., Gibson, J., Reid, G.A., Yonetani, T. & Schatz, G. (1982) Proc. Natl. Acad. Sci. USA *79*, 267-271.

30. Mihara, K., Omura, T., Harano, T., Brenner, S., Fleischer, S., Rajagopalan, K.V. & Blobel, G. (1982) J. Biol. Chem. *257*, 3355-3358.

31. Zimmermann, R., Paluch, U., Sprinzl, M. & Neupert, W. (1979) Eur. J. Biochem. *99*, 247-252.

32. Zimmermann, R. & Neupert, W. (1980) Eur. J. Biochem. *109*, 217-229.

33. Zwizinski, C., Schleyer, M. & Neupert, W. (1983) J. Biol. Chem. *258*, 4071-4074.

34. Schleyer, M. & Neupert, W. (1984) J. Biol. Chem. *259*, 3487-3491.

35. Schleyer, M., Schmidt, B. & Neupert, W. (1982) Eur. J. Biochem. *125*, 109-116.

36. Zwizinski, C., Schleyer, M. & Neupert, W. (1984) J. Biol. Chem. *259*, 7850-7856.

37. Maccecchini, M.L., Rudin, Y., Blobel, G. & Schatz, G. (1979) Proc. Natl. Acad. Sci. USA *76*, 343-347.

38. Gasser, S.M., Daum, G. & Schatz, G. (1982) J. Biol. Chem. *257*, 13034-13041.

39. Todd, R.D., McAda, P.C. & Douglas, M.G. (1979) J. Biol. Chem. *254*, 11134-11141.

40. Todd, R.D., Buck, M.A. & Douglas, M.G. (1981) J. Biol. Chem. *256*, 9037–9043.
41. Boehni, P.C., Gasser, S.M., Leaver, C. & Schatz, G. (1980) In The Organization and Expression of the Mitochondrial Genome, 423–433, Kroon, A.M. & Saccone, C., eds. Elsevier/North Holland, Amsterdam.
42. McAda, P.C. & Douglas, M.G. (1982) J. Biol. Chem. *257*, 3177–3182.
43. Lewin, A.S. & Norman, D.K. (1983) J. Biol. Chem. *258*, 6750–6755.
44. Reid, G.A. & Schatz, G. (1982) J. Biol. Chem. *257*, 13056–13061.
45. Reid, G.A. & Schatz, G. (1982) J. Biol. Chem. *257*, 13062–13067.
46. Zwizinski, C. & Neupert, W. (1983) J. Biol. Chem. *258*, 13340–13346.
47. Michel, R., Wachter, E. & Sebald, W. (1979) FEBS Lett. *101*, 373–376.
48. Viebrock, A., Perz, A. & Sebald, W. (1982) EMBO J. *1*, 565–571.
49. Schmidt, B., Hennig, B., Zimmermann, R. & Neupert, W. (1983) J. Cell Biol. *96*, 248–255.
50. Schmidt, B., Hennig, B., Koehler, H. & Neupert, W. (1983) J. Biol. Chem. *258*, 4687–4689.
51. Schmidt, B., Wachter, E., Sebald, W. & Neupert, W. (1984) Eur. J. Biochem. *144*, 581–588.
52. Miyazawa, S., Ozasa, H., Futura, S., Osumi, T., Hashimoto, T., Miura, S., Mori, M. & Tatibana, M. (1983) J. Biochem. *93*, 453–459.
53. Du Bois, R.N., Simpson, E.R., Tuckey, J., Lambeth, J.D. & Waterman, M.R. (1981) Proc. Natl. Acad. Sci. USA *78*, 1028–1032.
54. Lewin, A.S., Gregor, J., Mason, T.L., Nelson, N. & Schatz, G. (1980) Proc. Natl. Acad. Sci. USA *77*, 3998–4002.
55. Mihara, K. & Blobel, G. (1980) Proc. Natl. Acad. Sci. USA *77*, 4160–4164.
56. Lustig, A., Padmanaban, G. & Rabinowitz, M. (1982) Biochemistry *21*, 309–316.
57. Boehni, P.C., Daum, G. & Schatz, G. (1983) J. Biol. Chem. *258*, 4937–4943.
58. Cerletti, N., Boehni, P.C. & Suda, K. (1983) J. Biol. Chem. *258*, 4944–4949.
59. Schmelzer, E. & Heinrich, P.C. (1980) J. Biol. Chem. *255*, 7503–7506.
60. Schmelzer, E., Northemann, W., Kadenbach, B. & Heinrich, P.C. (1982) Eur. J. Biochem. *127*, 177–183.
61. Cote, C., Solioz, M. & Schatz, G. (1979) J. Biol. Chem. *254*, 1437–1439.
62. Ohashi, A., Gibson, J., Gregor, J. & Schatz, G. (1982) J. Biol. Chem. *257*, 13042–13047.
63. Van Loon, A.P.G.M., Kreike, J., De Ronde, A., vom Der Horst, T.J., Gasser, S.M. & Grivell, L.A. (1983) Eur. J. Biochem. *135*, 457–463.
64. Van Loon, A.P.G.M., De Groot, R.J., De Haan, M. Dekker, A., & Grivell, L.A. (1984) EMBO J. *3*, 1039–1043.
65. De Haan, M., Van Loon, A.P.G.M., Kreike, J., Vaessen, R.T.M. & Grivell, L.A. (1984) Eur. J. Biochem. *138*, 169–177.
66. Van Loon, A.P.G.M., Van Eijk, E. & Grivell, L.A. (1983) EMBO J. *2*, 1765–1770.
67. Van Loon, A.P.G.M., Maarse, A.C., Riezman, H. & Grivell, L.A. (1983) Gene. *26*, 261–272.
68. Teintze, M., Slaughter, M., Weiss, H. & Neupert, W. (1982) J. Biol. Chem. *257*, 10364–10371.
69. Ricquier, D., Thibault, J., Bouillaud, F. & Kuster, Y. (1983) J. Biol. Chem. *258*, 6675–6677.
70. Nabi, N. & Omura, T. (1980) Biochem. Biophys. Res. Commun. *97*, 680–686.
71. Yamauchi, K., Hayashi, N. & Kikuchi, G. (1980) FEBS Lett. *115*, 15–18.
72. Ades, J.Z. & Harpe, K.G. (1981) J. Biol. Chem. *256*, 9329–9333.

73. Yamauchi, K., Hayashi, N. & Kikuchi, G. (1980) J. Biol. Chem. *255*, 1746-1751.
74. Sonderegger, P., Jaussi, R. & Christen, P. (1980) Biochem. Biophys. Res. Commun. *94*, 1256-1260.
75. Sonderegger, P., Jaussi, R., Christen, P. & Gehring, H. (1982) J. Biol. Chem. *257*, 3339-3345.
76. Jaussi, R., Sonderegger, P., Fluckiger, J. & Christen, P. (1982) J. Biol. Chem. *257*, 13334-13340.
77. Shore, G.C., Carignan, P. & Raymond, Y. (1979) J. Biol. Chem. *254*, 3141-3144.
78. Mori, M., Morris, S.M., Jr. & Cohen, P.P. (1979) Proc. Natl. Acad. Sci. USA *76*, 3179-3183.
79. Mori, M., Miura, S., Tatibana, M. & Cohen, P.P. (1979) Proc. Natl. Acad. Sci. USA *76*, 5071-5075.
80. Raymond, Y. & Shore, G.C. (1979) J. Biol. Chem. *254*, 9335-9338.
81. Raymond, Y. & Shore, G.C. (1981) J. Biol. Chem. *256*, 2087-2090.
82. Campbell, M.T., Sutton, R. & Pollak, J.K. (1982) Eur. J. Biochem. *125*, 401-406.
83. Mori, M., Miura, S., Tatibana, M. & Cohen, P.P. (1981) J. Biol. Chem. *256*, 4127-4132.
84. Mori, M., Morita, T., Ikeda, F., Amaya, Y., Tatibana, M. & Cohen, P.P. (1981) Proc. Natl. Acad. Sci. USA *78*, 6056-6060.
85. Harmey, M.A. & Neupert, W. (1979) FEBS Lett. *108*, 385-389.
86. Miralles, V., Felipo, V., Hernandez-Yago, J. & Grisdia, S. (1982) Biochem. Biophys. Res. Commun. *107*, 1028-1036.
87. Felipo, V., Miralles, V., Knecht, G., Hernandez-Yago, J. & Grisolia, S. (1983) Eur. J. Biochem. *133*, 641-644.
88. Sakakibara, R., Kamisaki, Y. & Wada, H. (1981) Biochem. Biophys. Res. Commun. *102*, 235-242.
89. Hampsey, D.M., Lewin, A.S. & Kohlhaw, G.B. (1983) Proc. Natl. Acad. Sci. USA *80*, 1270-1274.
90. Mueckler, M.M., Himeno, M. & Pitot, H.C. (1982) J. Biol. Chem. *257*, 7178-7180.
91. Conboy, J.G., Kalousek, F. & Rosenberg, L.G. (1979) Proc. Natl. Acad. Sci. USA *76*, 5724-5727.
92. Mori, M., Miura, S., Tatibana, M. & Cohen, P.P. (1980) Proc. Natl. Acad. Sci. USA *77*, 7044-7048.
93. Conboy, J.G. & Rosenberg, L.E. (1981) Proc. Natl. Acad. Sci. USA *78*, 3073-3077.
94. Mori, M., Morita, T., Miura, S. & Tatibana, M. (1981) J. Biol. Chem. *256*, 8263-8266.
95. Kraus, J.P., Conboy, J.G. & Rosenberg, L.E. (1981) J. Biol. Chem. *256*, 10739-10742.
96. Conboy, J.G., Fenton, W.A. & Rosenberg, L.E. (1982) Biochem. Biophys. Res. Commun. *105*, 1-7.
97. Miura, S., Mori, M., Amaya, Y. & Tatibana, M. (1982) Eur. J. Biochem. *122*, 641-647.
98. Morita, T., Miura, M., Mori, M. & Tatibana, M. (1982) Eur. J. Biochem. *122*, 501-509.
99. Kolansky, D.M., Conboy, J.G., Fenton, W.A. & Rosenberg, L.E. (1982) J. Biol. Chem. *257*, 8467-8471.
100. Argan, C., Lusty, C.J. & Shore, G.C. (1983) J. Biol. Chem. *258*, 6667-6670.
101. Miura, S., Mori, M. & Tatibana, M. (1983) J. Biol. Chem. *258*, 6671-6674.

102. Lustig, A., Levens, D. & Rabinowitz, M. (1981) J. Biol. Chem. *257*, 5800–5808.
103. Autor, A.P. (1982) J. Biol. Chem. *257*, 2713–2718.
104. Anderson, L. (1981) Proc. Natl. Acad. Sci. USA *78*, 2407–2411.
105. Rietveld, A., Sijens, P., Verkleij, A.J. & de Kruijff, B. (1983) EMBO J. *2*, 907–913.
106. Date, T., Zwizinski, C., Ludmerer, S. & Wickner, W. (1980) Proc. Natl. Acad. Sci. USA *77*, 827–831.
107. Date, T., Goodman, J.M. & Wickner, W. (1980) Proc. Natl. Acad. Sci. USA *77*, 4469–4673.
108. Enequist, H.G., Hirst, T.R., Harayama, S., Hardy, S.J.S. & Randall, L.L. (1981) Eur. J. Biochem. *116*, 227–233.
109. Daniels, C.J., Bole, D.G., Quay, S.C. & Oxender, D.L. (1981) Proc. Natl. Acad. Sci. USA *78*, 5396–5400.
110. Zimmermann, R., Watts, C. & Wickner, W. (1982) J. Biol. Chem. *257*, 6529–6536.
111. Kruse, B. & Sebald, W. (1985) In H^+-ATPase (ATP Synthase): Structure, Function, Biogenesis: The F_0F_1 Complex of Coupling Membranes, Papa, S., Altendorf, K., Eruster, L. & Packer, L., eds. Adnatica Editrice, Ban.
112. Ohta, S. & Schatz, G. (1984) EMBO J. *3*, 651–657.
113. Nagata, S., Tsunetsugu-Yokota, Y., Naito, A. & Kaziro, Y. (1983) Proc. Natl. Acad. Sci. USA *80*, 6192–6196.
114. Maarse, A.C., Van Loon, A.P.G.M., Riezman, H., Gregor, I., Schatz, G. & Grivell, L.A. (1984) EMBO J. *3*, 2831–2837.
115. Morohashi, K., Fujii-Kuriyama, Y., Okada, Y., Sogawa, K., Hirose, T., Inayama, S. & Omura, T. (1984) Proc. Natl. Acad. Sci. USA *81*, 4647–4651.
116. Sadler, I., Suda, K., Schatz, G., Kaudewitz, F. & Haid, A. (1984) EMBO J. *3*, 2137–2143.
117. Wright, R.M., Ko, C., Cumsky, M.G. & Poyton, R.O. (1984) J. Biol. Chem. *259*, 15401–15407.
118. Takiguchi, M., Miura, S., Mori, M., Tatibana, M., Nagata, S. & Kaziro, Y. (1984) Proc. Natl. Acad. Sci. USA *81*, 7412–7416.
119. Harnisch, U., Weiss, H. & Sebald, W. (1985) Eur. J. Biochem. *149*, 95–99.
120. Nyunoya, H., Broglie, K.E., Widgren, E.E. & Lusty, C.J. (1985) J. Biol. Chem. *260*, 9346–9356.
121. Koerner, T.J., Hill, J. & Tzagoloff, A. (1985) J. Biol. Chem. *260*, 9513–9515.
122. Okamura, T., John, M.E., Zuber, M.X., Simpson, E.R. & Waterman, M.R. (1985) Proc. Natl. Acad. Sci. USA *82*, 5705–5709.
123. Joh, T., Nomiyama, H., Maeda, S., Shimada, K. & Moring, Y. (1985) Proc. Natl. Acad. Sci. USA *82*, 6065–6069.
124. Hurt, E.C., Pesold-Hurt, B. & Schatz, G. (1984) EMBO J. *3*, 3149–3156.
125. Hurt, E.C., Pesold-Hurt, B., Suda, K., Oppliger, W. & Schatz, G. (1985) EMBO J. *4*, 2061–2068.
126. Schleyer, M. & Neupert, W. (1985) Cell *43*, 339–350.
127. Pfaller, R., Freitag, H., Harmey, M.A., Benz, R. & Neupert, W. (1985) J. Biol. Chem. *260*, 8188–8193.
128. Yaffe, M.P. & Schatz, G. (1984) Proc. Natl. Acad. Sci. USA *81*, 4819–4823.
129. Yaffe, M.P., Ohta, S. & Schatz, G. (1985) EMBO J. *4*, 2069–2074.
130. Pfanner, N. & Neupert, W. (1985) EMBO J. *4*, 2819–2825.
131. Horwich, A.L., Kalousek, F., Mellman, L. & Rosenberg, L. (1985) EMBO J. *4*, 1129–1135.
132. Kellems, R.E., Allison, V.G. & Bntow, R.A. (1975) J. Cell Biol *65*, 1–14.

Index